THE NATURAL LANGUAGE PROCESSING
WORKSHOP

Confidently design and build your own NLP projects
with this easy-to-understand practical guide

Rohan Chopra, Aniruddha M. Godbole, Nipun Sadvilkar,
Muzaffar Bashir Shah, Sohom Ghosh, and Dwight Gunning

THE NATURAL LANGUAGE PROCESSING WORKSHOP

Authors: Rohan Chopra, Aniruddha M. Godbole, Nipun Sadvilkar, Muzaffar Bashir Shah, Sohom Ghosh, and Dwight Gunning

Reviewers: Ankit Bhatia, Nagendra Nagaraj, Nimish Narang, Sumit Kumar Raj, Tom Taulli, and Ankit Verma

Managing Editor: Saumya Jha

Acquisitions Editors: Royluis Rodrigues, Kunal Sawant, Sneha Shinde, Archie Vankar, and Karan Wadekar

Production Editor: Roshan Kawale

Editorial Board: Megan Carlisle, Samuel Christa, Mahesh Dhyani, Heather Gopsill, Manasa Kumar, Alex Mazonowicz, Monesh Mirpuri, Bridget Neale, Dominic Pereira, Shiny Poojary, Abhishek Rane, Brendan Rodrigues, Erol Staveley, Ankita Thakur, Nitesh Thakur, and Jonathan Wray

First published: August 2020

Production reference: 2020321

ISBN: 978-1-80020-842-1

Published by Packt Publishing Ltd.

Livery Place, 35 Livery Street

Birmingham B3 2PB, UK

WHY LEARN WITH A PACKT WORKSHOP?

LEARN BY DOING

Packt Workshops are built around the idea that the best way to learn something new is by getting hands-on experience. We know that learning a language or technology isn't just an academic pursuit. It's a journey towards the effective use of a new tool—whether that's to kickstart your career, automate repetitive tasks, or just build some cool stuff.

That's why Workshops are designed to get you writing code from the very beginning. You'll start fairly small—learning how to implement some basic functionality—but once you've completed that, you'll have the confidence and understanding to move onto something slightly more advanced.

As you work through each chapter, you'll build your understanding in a coherent, logical way, adding new skills to your toolkit and working on increasingly complex and challenging problems.

CONTEXT IS KEY

All new concepts are introduced in the context of realistic use-cases, and then demonstrated practically with guided exercises. At the end of each chapter, you'll find an activity that challenges you to draw together what you've learned and apply your new skills to solve a problem or build something new.

We believe this is the most effective way of building your understanding and confidence. Experiencing real applications of the code will help you get used to the syntax and see how the tools and techniques are applied in real projects.

BUILD REAL-WORLD UNDERSTANDING

Of course, you do need some theory. But unlike many tutorials, which force you to wade through pages and pages of dry technical explanations and assume too much prior knowledge, Workshops only tell you what you actually need to know to be able to get started making things. Explanations are clear, simple, and to-the-point. So you don't need to worry about how everything works under the hood; you can just get on and use it.

Written by industry professionals, you'll see how concepts are relevant to real-world work, helping to get you beyond "Hello, world!" and build relevant, productive skills. Whether you're studying web development, data science, or a core programming language, you'll start to think like a problem solver and build your understanding and confidence through contextual, targeted practice.

ENJOY THE JOURNEY

Learning something new is a journey from where you are now to where you want to be, and this Workshop is just a vehicle to get you there. We hope that you find it to be a productive and enjoyable learning experience.

Packt has a wide range of different Workshops available, covering the following topic areas:

- Programming languages

- Web development

- Data science, machine learning, and artificial intelligence

- Containers

Once you've worked your way through this Workshop, why not continue your journey with another? You can find the full range online at http://packt.live/2MNkuyl.

If you could leave us a review while you're there, that would be great. We value all feedback. It helps us to continually improve and make better books for our readers, and also helps prospective customers make an informed decision about their purchase.

Thank you,
The Packt Workshop Team

Table of Contents

Chapter 2: Feature Extraction Methods 35

Chapter 3: Developing a Text Classifier 107

Chapter 5: Topic Modeling

Chapter 6: Vector Representation 271

Chapter 7: Text Generation and Summarization 319

Chapter 8: Sentiment Analysis   343

PREFACE

ABOUT THE BOOK

Do you want to learn how to communicate with computer systems using Natural Language Processing (NLP) techniques, or make a machine understand human sentiments? Do you want to build applications like Siri, Alexa, or chatbots, even if you've never done it before?

With *The Natural Language Processing Workshop*, you can expect to make consistent progress as a beginner, and get up to speed in an interactive way, with the help of hands-on activities and fun exercises.

The book starts with an introduction to NLP. You'll study different approaches to NLP tasks, and perform exercises in Python to understand the process of preparing datasets for NLP models. Next, you'll use advanced NLP algorithms and visualization techniques to collect datasets from open websites, and to summarize and generate random text from a document. In the final chapters, you'll use NLP to create a chatbot that detects positive or negative sentiment in text documents such as movie reviews.

By the end of this book, you'll be equipped with the essential NLP tools and techniques you need to solve common business problems that involve processing text.

AUDIENCE

This book is for beginner to mid-level data scientists, machine learning developers, and NLP enthusiasts. A basic understanding of machine learning and NLP is required to help you grasp the topics in this workshop more quickly.

ABOUT THE CHAPTERS

Chapter 1, Introduction to Natural Language Processing, starts by defining natural language processing and the different types of natural language processing tasks, using practical examples for each type. This chapter also covers the process of structuring and implementing a natural language processing project.

Chapter 2, Feature Extraction Methods, covers basic feature extraction methods from unstructured text. These include tokenization, stemming, lemmatization, and stopword removal. We also discuss observations we might see from these extraction methods and introduce Zipf's Law. Finally, we discuss the Bag of Words model and Term Frequency-Inverse Document Frequency (TF-IDF).

Chapter 3, Developing a Text Classifier, teaches you how to create a simple text classifier with feature extraction methods covered in the previous chapters.

Chapter 4, Collecting Text Data with Web Scraping and APIs, introduces you to web scraping and discusses various methods of collecting and processing text data from online sources, such as HTML and XML files and APIs.

Chapter 5, Topic Modeling, introduces topic modeling, an unsupervised natural language processing technique that groups documents according to topic. You will see how this is done using Latent Dirichlet Allocation (LDA), Latent Semantic Analysis (LSA), and Hierarchical Dirichlet Processes (HDP).

Chapter 6, Vector Representation, discusses the importance of representing text as vectors, and various vector representations, such as Word2Vec and Doc2Vec.

Chapter 7, Text Generation and Summarization, teaches you two simple natural language processing tasks: creating text summaries and generating random text with statistical assumptions and algorithms.

Chapter 8, Sentiment Analysis, teaches you how to detect sentiment in text, using simple techniques. Sentiment analysis is the use of computer algorithms to detect whether the sentiment of text is positive or negative.

CONVENTIONS

Code words in text, database table names, folder names, filenames, file extensions, pathnames, dummy URLs, user input, and Twitter handles are shown as follows: "We find that the summary for the Wikipedia article is much more coherent than the short story. We can also see that the summary with a **ratio** of **0.20** is a subset of a summary with a **ratio** of **0.25**."

Words that you see on the screen, for example, in menus or dialog boxes, also appear in the text like this: "On this page, click on **Keys** option to access the secret keys."

A block of code is set as follows:

```
text_after_twenty=text_after_twenty.replace('\n',' ')
text_after_twenty=re.sub(r"\s+"," ",text_after_twenty)
```

New terms and important words are shown like this: "A **Markov chain** consists of a state space and a specific type of successor function."

Long code snippets are truncated and the corresponding names of the code files on GitHub are placed at the top of the truncated code. The permalinks to the entire code are placed below the code snippet. It should look as follows:

Exercise 7.01.ipynb

```
1   HANDLE = '@\w+\n'
2   LINK = 'https?://t\.co/\w+'
3   SPECIAL_CHARS = '&lt;|&lt;|&amp;|#'
4   PARA='\n+'
5   def clean(text):
6       #text = re.sub(HANDLE, ' ', text)
7       text = re.sub(LINK, ' ', text)
8       text = re.sub(SPECIAL_CHARS, ' ', text)
9       text = re.sub(PARA, '\n', text)
```

The full code can be found at https://packt.live/2D7RPPZ.

CODE PRESENTATION

Lines of code that span multiple lines are split using a backslash (\ ). When the code is executed, Python will ignore the backslash, and treat the code on the next line as a direct continuation of the current line.

For example:

```
history = model.fit(X, y, epochs=100, batch_size=5, verbose=1, \
                    validation_split=0.2, shuffle=False)
```

Comments are added into code to help explain specific bits of logic. Single-line comments are denoted using the **#** symbol, as follows:

```
# Print the sizes of the dataset
print("Number of Examples in the Dataset = ", X.shape[0])
print("Number of Features for each example = ", X.shape[1])
```

Multi-line comments are enclosed by triple quotes, as shown below:

```
"""
Define a seed for the random number generator to ensure the
result will be reproducible
"""
seed = 1
np.random.seed(seed)
random.set_seed(seed)
```

SETTING UP YOUR ENVIRONMENT

Before we explore the book in detail, we need to set up specific software and tools. In the following section, we shall see how to do that.

INSTALLATION AND SETUP

Jupyter notebooks are available once you install Anaconda on your system. Anaconda can be installed for Windows systems using the steps available at https://docs.anaconda.com/anaconda/install/windows/.

For other systems, navigate to the respective installation guide from https://docs.anaconda.com/anaconda/install/.

These installations will be executed in the C drive of your system. You can choose to change the destination.

INSTALLING THE REQUIRED LIBRARIES

Open Anaconda Prompt and follow the steps given here to get your system ready. We will create a new environment on Anaconda where we will install all the required libraries and run our code:

1. To create a new environment, run the following command:

```
conda create --name nlp
```

2. To activate the environment, type the following:

```
conda activate nlp
```

For this course, whenever you are asked to open a terminal, you need to open Anaconda Prompt, activate the environment, and then proceed.

3. To install all the libraries, download the environment file from https://packt.live/30qfL9V and run the following command:

```
pip install -f requirements.txt
```

4. Jupyter notebooks allow us to run code and experiment with code blocks. To start Jupyter Notebook, run the following inside the **nlp** environment:

```
jupyter notebook
```

A new browser window will open up with the Jupyter interface. You can now navigate to the project location and run Jupyter Notebook.

INSTALLING LIBRARIES

`pip` comes pre-installed with Anaconda. Once Anaconda is installed on your machine, all the required libraries can be installed using `pip`, for example, `pip install numpy`. Alternatively, you can install all the required libraries using `pip install -r requirements.txt`. You can find the `requirements.txt` file at https://packt.live/39RZuOh.

The exercises and activities will be executed in Jupyter Notebooks. Jupyter is a Python library and can be installed in the same way as the other Python libraries – that is, with `pip install jupyter`, but fortunately, it comes pre-installed with Anaconda. To open a notebook, simply run the command `jupyter notebook` in the Terminal or Command Prompt.

ACCESSING THE CODE FILES

You can find the complete code files of this book at https://packt.live/3fJ4qap. You can also run many activities and exercises directly in your web browser by using the interactive lab environment at https://packt.live/3gwk4WQ.

We've tried to support interactive versions of all activities and exercises, but we recommend a local installation as well for instances where this support isn't available.

If you have any issues or questions about installation, please email us at `workshops@packt.com`.

1

INTRODUCTION TO NATURAL LANGUAGE PROCESSING

OVERVIEW

In this chapter, you will learn the difference between **Natural Language Processing** (**NLP**) and basic text analytics. You will implement various preprocessing tasks such as tokenization, lemmatization, stemming, stop word removal, and more. By the end of this chapter, you will have a deep understanding of the various phases of an NLP project, from data collection to model deployment.

INTRODUCTION

Before we can get into NLP in any depth, we first need to understand what natural language is. To put it in simple terms, it is a means for us to express our thoughts and ideas. To define it more specifically, language is a mutually agreed upon set of protocols involving words/sounds that we use to communicate with each other.

In this era of digitization and computation, we are constantly interacting with machines around us through various means, such as voice commands and typing instructions in the form of words. Thus, it has become essential to develop mechanisms by which human language can be comprehended accurately by computers. NLP helps us do this. So, NLP can be defined as a field of computer science that is concerned with enabling computer algorithms to understand, analyze, and generate natural languages.

Let's look at an example. You have probably interacted with Siri or Alexa at some point. Ask Alexa for a cricket score, and it will reply with the current score. The technology behind this is NLP. Siri and Alexa use techniques such as Speech to Text with the help of a search engine to do this magic. As the name suggests, Speech to Text is an application of NLP in which computers are trained to understand verbally spoken words.

NLP works at different levels, which means that machines process and understand natural language at different levels. These levels are as follows:

- **Morphological level**: This level deals with understanding word structure and word information.

- **Lexical level**: This level deals with understanding the part of speech of the word.

- **Syntactic level**: This level deals with understanding the syntactic analysis of a sentence, or parsing a sentence.

- **Semantic level**: This level deals with understanding the actual meaning of a sentence.

- **Discourse level**: This level deals with understanding the meaning of a sentence beyond just the sentence level, that is, considering the context.

- **Pragmatic level**: This level deals with using real-world knowledge to understand the sentence.

HISTORY OF NLP

NLP is a field that has emerged from various other fields such as artificial intelligence, linguistics, and data science. With the advancement of computing technologies and the increased availability of data, NLP has undergone a huge change. Previously, a traditional rule-based system was used for computations, in which you had to explicitly write hardcoded rules. Today, computations on natural language are being done using machine learning and deep learning techniques.

Consider an example. Let's say we have to extract the names of some politicians from a set of political news articles. So, if we want to apply rule-based grammar, we must manually craft certain rules based on human understanding of language. Some of the rules for extracting a person's name can be that the word should be a proper noun, every word should start with a capital letter, and so on. As we can see, using a rule-based system like this would not yield very accurate results.

Rule-based systems do work well in some cases, but the disadvantages far outweigh the advantages. One major disadvantage is that the same rule cannot be applicable in all cases, given the complex and nuanced nature of most language. These disadvantages can be overcome by using machine learning, where we write an algorithm that tries to learn a language using the text corpus (training data) rather than us explicitly programming it to do so.

TEXT ANALYTICS AND NLP

Text analytics is the method of extracting meaningful insights and answering questions from text data, such as those to do with the length of sentences, length of words, word count, and finding words from the text. Let's understand this with an example.

Suppose we are doing a survey using news articles. Let's say we have to find the top five countries that contributed the most in the field of space technology in the past 5 years. So, we will collect all the space technology-related news from the past 5 years using the Google News API. Now, we must extract the names of countries in these news articles. We can perform this task using a file containing a list of all the countries in the world.

Next, we will create a dictionary in which keys will be the country names and their values will be the number of times the country name is found in the news articles. To search for a country in the news articles, we can use a simple word regex. After we have completed searching all the news articles, we can sort the country names by the values associated with them. In this way, we will come up with the top five countries that contributed the most to space technology in the last 5 years.

This is a typical example of text analytics, in which we are generating insights from text without getting into the semantics of the language.

It is important here to note the difference between text analytics and NLP. The art of extracting useful insights from any given text data can be referred to as text analytics. NLP, on the other hand, helps us in understanding the semantics and the underlying meaning of text, such as the sentiment of a sentence, top keywords in text, and parts of speech for different words. It is not just restricted to text data; voice (speech) recognition and analysis also come under the domain of NLP. It can be broadly categorized into two types: **Natural Language Understanding (NLU)** and **Natural Language Generation (NLG)**. A proper explanation of these terms is provided here:

- **NLU**: NLU refers to a process by which an inanimate object with computing power is able to comprehend spoken language. As mentioned earlier, Siri and Alexa use techniques such as Speech to Text to answer different questions, including inquiries about the weather, the latest news updates, live match scores, and more.

- **NLG**: NLG refers to a process by which an inanimate object with computing power is able to communicate with humans in a language that they can understand or is able to generate human-understandable text from a dataset. Continuing with the example of Siri or Alexa, ask one of them about the chances of rainfall in your city. It will reply with something along the lines of, "Currently, there is no chance of rainfall in your city." It gets the answer to your query from different sources using a search engine and then summarizes the results. Then, it uses Text to Speech to relay the results in verbally spoken words.

So, when a human speaks to a machine, the machine interprets the language with the help of the NLU process. By using the NLG process, the machine generates an appropriate response and shares it with the human, thus making it easier for humans to understand the machine. These tasks, which are part of NLP, are not part of text analytics. Let's walk through the basics of text analytics and see how we can execute it in Python.

Before going to the exercises, let's define some prerequisites for running the exercises. Whether you are using Windows, Mac or Linux, you need to run your Jupyter Notebook in a virtual environment. You will also need to ensure that you have installed the requirements as stated in the *requirements.txt* file on https://packt.live/3fJ4qap.

EXERCISE 1.01: BASIC TEXT ANALYTICS

In this exercise, we will perform some basic text analytics on some given text data, including searching for a particular word, finding the index of a word, and finding a word at a given position. Follow these steps to implement this exercise using the following sentence:

"The quick brown fox jumps over the lazy dog."

1. Open a Jupyter Notebook.

2. Assign a **sentence** variable the value **'The quick brown fox jumps over the lazy dog'**. Insert a new cell and add the following code to implement this:

```
sentence = 'The quick brown fox jumps over the lazy dog'
sentence
```

3. Check whether the word **'quick'** belongs to that text using the following code:

```
def find_word(word, sentence):
    return word in sentence
find_word('quick', sentence)
```

The preceding code will return the output **'True'**.

4. Find out the **index** value of the word **'fox'** using the following code:

```
def get_index(word, text):
    return text.index(word)
get_index('fox', sentence)
```

The code will return the output **16**.

5. To find out the rank of the word **'lazy'**, use the following code:

```
get_index('lazy', sentence.split())
```

This code generates the output **7**.

6. To print the third word of the given text, use the following code:

```
def get_word(text, rank):
    return text.split()[rank]
get_word(sentence, 2)
```

This will return the output **brown**.

7. To print the third word of the given sentence in reverse order, use the following code:

```
get_word(sentence, 2)[::-1]
```

This will return the output **nworb**.

8. To concatenate the first and last words of the given sentence, use the following code:

```
def concat_words(text):
    """
    This method will concat first and last
    words of given text
    """
    words = text.split()
    first_word = words[0]
    last_word = words[len(words)-1]
    return first_word + last_word
concat_words(sentence)
```

> **NOTE**
>
> The triple-quotes (**"""**) shown in the code snippet above are used to denote the start and end points of a multi-line code comment. Comments are added into code to help explain specific bits of logic.

The code will generate the output **Thedog**.

9. To print words at even positions, use the following code:

```
def get_even_position_words(text):
    words = text.split()
    return [words[i] for i in range(len(words)) if i%2 == 0]


get_even_position_words(sentence)
```

This code generates the following output:

```
['The', 'brown', 'jumps', 'the', 'dog']
```

10. To print the last three letters of the text, use the following code:

```
def get_last_n_letters(text, n):
    return text[-n:]
get_last_n_letters(sentence,3)
```

This will generate the output **dog**.

11. To print the text in reverse order, use the following code:

```
def get_reverse(text):
    return text[::-1]
get_reverse(sentence)
```

This code generates the following output:

```
'god yzal eht revo spmuj xof nworb kciuq ehT'
```

12. To print each word of the given text in reverse order, maintaining their sequence, use the following code:

```
def get_word_reverse(text):
    words = text.split()
    return ' '.join([word[::-1] for word in words])
get_word_reverse(sentence)
```

This code generates the following output:

```
ehT kciuq nworb xof spmuj revo eht yzal god
```

We are now well acquainted with basic text analytics techniques.

> **NOTE**
>
> To access the source code for this specific section, please refer to https://packt.live/38Yrf77.
>
> You can also run this example online at https://packt.live/2ZsCvpf.

In the next section, let's dive deeper into the various steps and subtasks in NLP.

VARIOUS STEPS IN NLP

We've talked about the types of computations that are done with natural language. Apart from these basic tasks, you can also design your own tasks as per your requirements. In the coming sections, we will discuss the various preprocessing tasks in detail and demonstrate each of them with an exercise.

To perform these tasks, we will be using a Python library called **NLTK** (**Natural Language Toolkit**). NLTK is a powerful open source tool that provides a set of methods and algorithms to perform a wide range of NLP tasks, including tokenizing, parts-of-speech tagging, stemming, lemmatization, and more.

TOKENIZATION

Tokenization refers to the procedure of splitting a sentence into its constituent parts—the words and punctuation that it is made up of. It is different from simply splitting the sentence on whitespaces, and instead actually divides the sentence into constituent words, numbers (if any), and punctuation, which may not always be separated by whitespaces. For example, consider this sentence: "I am reading a book." Here, our task is to extract words/tokens from this sentence. After passing this sentence to a tokenization program, the extracted words/tokens would be "I," "am," "reading," "a," "book," and "." – this example extracts one token at a time. Such tokens are called **unigrams**.

NLTK provides a method called **word_tokenize()**, which tokenizes given text into words. It actually separates the text into different words based on punctuation and spaces between words.

To get a better understanding of tokenization, let's solve an exercise based on it in the next section.

EXERCISE 1.02: TOKENIZATION OF A SIMPLE SENTENCE

In this exercise, we will tokenize the words in a given sentence with the help of the **NLTK** library. Follow these steps to implement this exercise using the sentence, "I am reading NLP Fundamentals."

1. Open a Jupyter Notebook.

2. Insert a new cell and add the following code to import the necessary libraries and download the different types of NLTK data that we are going to use for different tasks in the following exercises:

```
from nltk import word_tokenize, download
download(['punkt','averaged_perceptron_tagger','stopwords'])
```

    In the preceding code, we are using NLTK's **download()** method, which downloads the given data from NLTK. NLTK data contains different corpora and trained models. In the preceding example, we will be downloading the stop word list, **'punkt'**, and a perceptron tagger, which is used to implement parts of speech tagging using a structured algorithm. The data will be downloaded at **nltk_data/corpora/** in the home directory of your computer. Then, it will be loaded from the same path in further steps.

3. The **word_tokenize()** method is used to split the sentence into words/ tokens. We need to add a sentence as input to the **word_tokenize()** method so that it performs its job. The result obtained will be a list, which we will store in a **word** variable. To implement this, insert a new cell and add the following code:

```
def get_tokens(sentence):
    words = word_tokenize(sentence)
    return words
```

4. In order to view the list of tokens generated, we need to view it using the **print()** function. Insert a new cell and add the following code to implement this:

```
print(get_tokens("I am reading NLP Fundamentals."))
```

    This code generates the following output:

```
['I', 'am', 'reading', 'NLP', 'Fundamentals', '.']
```

We can see the list of tokens generated with the help of the
word_tokenize() method.

> **NOTE**
>
> To access the source code for this specific section, please refer
> to https://packt.live/30bGG85.
>
> You can also run this example online at https://packt.live/30dK1mZ.

In the next section, we will see another pre-processing step:
Parts-of-Speech (PoS) tagging.

POS TAGGING

In NLP, the term PoS refers to parts of speech. PoS tagging refers to the process
of tagging words within sentences with their respective PoS. We extract the PoS of
tokens constituting a sentence so that we can filter out the PoS that are of interest
and analyze them. For example, if we look at the sentence, "The sky is blue," we get
four tokens, namely "The," "sky," "is," and "blue", with the help of tokenization. Now,
using a **PoS tagger**, we tag the PoS for each word/token. This will look as follows:

```
[('The', 'DT'), ('sky', 'NN'), ('is', 'VBZ'), ('blue', 'JJ')]
```

The preceding format is an output of the NLTK **pos_tag()** method. It is a list of
tuples in which every tuple consists of the word followed by the PoS tag:

DT = Determiner

NN = Noun, common, singular or mass

VBZ = Verb, present tense, third-person singular

JJ = Adjective

For the complete list of PoS tags in NLTK, you can refer
to https://pythonprogramming.net/natural-language-toolkit-nltk-part-speech-tagging/.

PoS tagging is performed using different techniques, one of which is a rule-based
approach that builds a list to assign a possible tag for each word.

PoS tagging finds application in many NLP tasks, including word sense disambiguation, classification, **Named Entity Recognition** (**NER**), and coreference resolution. For example, consider the usage of the word "planted" in these two sentences: "He planted the evidence for the case " and " He planted five trees in the garden. " We can see that the PoS tag of "planted" would clearly help us in differentiating between the different meanings of the sentences.

Let's perform a simple exercise to understand how PoS tagging is done in Python.

EXERCISE 1.03: POS TAGGING

In this exercise, we will find out the PoS for each word in the sentence, **I am reading NLP Fundamentals**. We first make use of tokenization in order to get the tokens. Later, we will use the **pos_tag()** method, which will help us find the PoS for each word/token. Follow these steps to implement this exercise:

1. Open a Jupyter Notebook.

2. Insert a new cell and add the following code to import the necessary libraries:

    ```
    from nltk import word_tokenize, pos_tag
    ```

3. To find the tokens in the sentence, we make use of the **word_tokenize()** method. Insert a new cell and add the following code to implement this:

    ```
    def get_tokens(sentence):
        words = word_tokenize(sentence)
        return words
    ```

4. Print the tokens with the help of the **print()** function. To implement this, add a new cell and write the following code:

    ```
    words  = get_tokens("I am reading NLP Fundamentals")

    print(words)
    ```

    This code generates the following output:

    ```
    ['I', 'am', 'reading', 'NLP', 'Fundamentals']
    ```

5. We'll now use the **pos_tag()** method. Insert a new cell and add the following code:

```
def get_pos(words):
    return pos_tag(words)
get_pos(words)
```

This code generates the following output:

```
[('I', 'PRP'),
 ('am', 'VBP'),
 ('reading', 'VBG'),
 ('NLP', 'NNP'),
 ('Fundamentals', 'NNS')]
```

In the preceding output, we can see that for each token, a PoS has been allotted. Here, **PRP** stands for **personal pronoun**, **VBP** stands for **verb present**, **VGB** stands for **verb gerund**, **NNP** stands for **proper noun singular**, and **NNS** stands for **noun plural**.

> **NOTE**
>
> To access the source code for this specific section, please refer to https://packt.live/306WY24.
>
> You can also run this example online at https://packt.live/38VLDpF.

We have learned about assigning appropriate PoS labels to tokens in a sentence. In the next section, we will learn about **stop words** in sentences and ways to deal with them.

STOP WORD REMOVAL

Stop words are the most frequently occurring words in any language and they are just used to support the construction of sentences and do not contribute anything to the semantics of a sentence. So, we can remove stop words from any text before an NLP process, as they occur very frequently and their presence doesn't have much impact on the sense of a sentence. Removing them will help us clean our data, making its analysis much more efficient. Examples of stop words include "a," "am," "and," "the," "in," "of," and more.

In the next exercise, we will look at the practical implementation of removing stop words from a given sentence.

EXERCISE 1.04: STOP WORD REMOVAL

In this exercise, we will check the list of stop words provided by the **nltk** library. Based on this list, we will filter out the stop words included in our text:

1. Open a Jupyter Notebook.

2. Insert a new cell and add the following code to import the necessary libraries:

```
from nltk import download
download('stopwords')
from nltk import word_tokenize
from nltk.corpus import stopwords
```

3. In order to check the list of stop words provided for **English**, we pass it as a parameter to the **words()** function. Insert a new cell and add the following code to implement this:

```
stop_words = stopwords.words('english')
```

4. In the code, the list of stop words provided by **English** is stored in the **stop_words** variable. In order to view the list, we make use of the **print()** function. Insert a new cell and add the following code to view the list:

```
print(stop_words)
```

This code generates the following output:

```
['i', 'me', 'my', 'myself', 'we', 'our', 'ours', 'ourselves', 'you', "you're", "you've",
"you'll", "you'd", 'your', 'yours', 'yourself', 'yourselves', 'he', 'him', 'his', 'himsel
f', 'she', "she's", 'her', 'hers', 'herself', 'it', "it's", 'its', 'itself', 'they', 'the
m', 'their', 'theirs', 'themselves', 'what', 'which', 'who', 'whom', 'this', 'that', "tha
t'll", 'these', 'those', 'am', 'is', 'are', 'was', 'were', 'be', 'been', 'being', 'have',
'has', 'had', 'having', 'do', 'does', 'did', 'doing', 'a', 'an', 'the', 'and', 'but', 'i
f', 'or', 'because', 'as', 'until', 'while', 'of', 'at', 'by', 'for', 'with', 'about', 'ag
ainst', 'between', 'into', 'through', 'during', 'before', 'after', 'above', 'below', 'to',
'from', 'up', 'down', 'in', 'out', 'on', 'off', 'over', 'under', 'again', 'further', 'the
n', 'once', 'here', 'there', 'when', 'where', 'why', 'how', 'all', 'any', 'both', 'each',
'few', 'more', 'most', 'other', 'some', 'such', 'no', 'nor', 'not', 'only', 'own', 'same',
'so', 'than', 'too', 'very', 's', 't', 'can', 'will', 'just', 'don', "don't", 'should', "s
hould've", 'now', 'd', 'll', 'm', 'o', 're', 've', 'y', 'ain', 'aren', "aren't", 'couldn',
"couldn't", 'didn', "didn't", 'doesn', "doesn't", 'hadn', "hadn't", 'hasn', "hasn't", 'hav
en', "haven't", 'isn', "isn't", 'ma', 'mightn', "mightn't", 'mustn', "mustn't", 'needn',
"needn't", 'shan', "shan't", 'shouldn', "shouldn't", 'wasn', "wasn't", 'weren', "weren't",
'won', "won't", 'wouldn', "wouldn't"]
```

Figure 1.1: List of stop words provided by English

5. To remove the stop words from a sentence, we first assign a string to the **sentence** variable and tokenize it into words using the **word_tokenize()** method. Insert a new cell and add the following code to implement this:

```
sentence = "I am learning Python. It is one of the "\
           "most popular programming languages"
sentence_words = word_tokenize(sentence)
```

> **NOTE**
>
> The code snippet shown here uses a backslash (\ ) to split the logic across multiple lines. When the code is executed, Python will ignore the backslash, and treat the code on the next line as a direct continuation of the current line.

6. To print the list of tokens, insert a new cell and add the following code:

```
print(sentence_words)
```

This code generates the following output:

```
['I', 'am', 'learning', 'Python', '.', 'It', 'is', 'one', 'of',
'the', 'most', 'popular', 'programming', 'languages']
```

7. To remove the stop words, we need to loop through each word in the sentence, check whether there are any stop words, and then finally combine them to form a complete sentence. To implement this, insert a new cell and add the following code:

```
def remove_stop_words(sentence_words, stop_words):
    return ' '.join([word for word in sentence_words if \
                    word not in stop_words])
```

8. To check whether the stop words are filtered out from our sentence, print the **sentence_no_stops** variable. Insert a new cell and add the following code to print:

```
print(remove_stop_words(sentence_words,stop_words))
```

This code generates the following output:

```
I learning Python. It one popular programming languages
```

As you can see in the preceding code snippet, stop words such as "am," "is," "of," "the," and "most" are being filtered out and text without stop words is produced as output.

9. Add your own stop words to the stop word list:

```
stop_words.extend(['I','It', 'one'])
print(remove_stop_words(sentence_words,stop_words))
```

This code generates the following output:

```
learning Python . popular programming languages
```

As we can see from the output, now words such as "I," "It," and* "One" are removed as we have added them to our custom stop word list. We have learned how to remove stop words from given text.

> ## NOTE
>
> To access the source code for this specific section, please refer to https://packt.live/3j4KBw7.
>
> You can also run this example online at https://packt.live/3fyYSir.

In the next section, we will focus on normalizing text.

TEXT NORMALIZATION

There are some words that are spelled, pronounced, and represented differently—for example, words such as Mumbai and Bombay, and US and United States. Although they are different, they refer to the same thing. There are also different forms of words that need to be converted into base forms. For example, words such as "does" and "doing," when converted to their base form, become "do." Along these lines, **text normalization** is a process wherein different variations of text get converted into a standard form. We need to perform text normalization as there are some words that can mean the same thing as each other. There are various ways of normalizing text, such as spelling correction, stemming, and lemmatization, which will be covered later.

For a better understanding of this topic, we will look into a practical implementation of text normalization in the next section.

EXERCISE 1.05: TEXT NORMALIZATION

In this exercise, we will normalize some given text. Basically, we will be trying to replace select words with new words, using the **replace()** function, and finally produce the normalized text. **replace()** is a built-in Python function that works on strings and takes two arguments. It will return a copy of a string in which the occurrence of the first argument will be replaced by the second argument.

Follow these steps to complete this exercise:

1. Open a Jupyter Notebook.

2. Insert a new cell and add the following code to assign a string to the **sentence** variable:

```
sentence = "I visited the US from the UK on 22-10-18"
```

3. We want to replace **"US"** with **"United States"**, **"UK"** with **"United Kingdom"**, and **"18"** with **"2018"**. To do so, use the **replace()** function and store the updated output in the **"normalized_sentence"** variable. Insert a new cell and add the following code to implement this:

```
def normalize(text):
    return text.replace("US", "United States")\
               .replace("UK", "United Kingdom")\
               .replace("-18", "-2018")
```

4. To check whether the text has been normalized, insert a new cell and add the following code to print it:

```
normalized_sentence = normalize(sentence)
print(normalized_sentence)
```

The code generates the following output:

```
I visited the United States from the United Kingdom on 22-10-2018
```

5. Add the following code:

```
normalized_sentence = normalize('US and UK are two superpowers')
print(normalized_sentence)
```

The code generates following output:

```
United States and United Kingdom are two superpowers
```

In the preceding code, we can see that our text has been normalized.

> **NOTE**
>
> To access the source code for this specific section, please refer to https://packt.live/2Wm49T8.
>
> You can also run this example online at https://packt.live/2Wm4d5k.

Over the next sections, we will explore various other ways in which text can be normalized.

SPELLING CORRECTION

Spelling correction is one of the most important tasks in any NLP project. It can be time-consuming, but without it, there are high chances of losing out on important information.

Spelling correction is executed in two steps:

1. Identify the misspelled word, which can be done by a simple dictionary lookup. If there is no match found in the language dictionary, it is considered to be misspelled.

2. Replace it or suggest the correctly spelled word. There are a lot of algorithms for this task. One of them is the minimum edit distance algorithm, which chooses the nearest correctly spelled word for a misspelled word. The nearness is defined by the number of edits that need to be made to the misspelled word to reach the correctly spelled word. For example, let's say there is a misspelled word, "autocorect." Now, to make it "autocorrect," we need to add one "r," and to make it "auto," we need to delete 6 characters, which means that "autocorrect" is the correct spelling because it requires the fewest edits.

We make use of the **autocorrect** Python library to correct spellings.

autocorrect is a Python library used to correct the spelling of misspelled words for different languages. It provides a method called **spell()**, which takes a word as input and returns the correct spelling of the word.

Let's look at the following exercise to get a better understanding of this.

EXERCISE 1.06: SPELLING CORRECTION OF A WORD AND A SENTENCE

In this exercise, we will perform spelling correction on a word and a sentence, with the help of Python's **autocorrect** library. Follow these steps in order to complete this exercise:

1. Open a Jupyter Notebook.

2. Insert a new cell and add the following code to import the necessary libraries:

```
from nltk import word_tokenize
from autocorrect import Speller
```

3. In order to correct the spelling of a word, pass a wrongly spelled word as a parameter to the **spell()** function. Before that, you have to create a **spell** object of the **Speller** class using **lang='en'** to signify the English language. Insert a new cell and add the following code to implement this:

```
spell = Speller(lang='en')
spell('Natureal')
```

This code generates the following output:

```
'Natural'
```

4. To correct the spelling of a sentence, first tokenize it into tokens. After that, loop through each token in **sentence**, autocorrect the words, and finally combine the words. Insert a new cell and add the following code to implement this:

```
sentence = word_tokenize("Ntural Luanguage Processin deals with "\
                         "the art of extracting insightes from "\
                         "Natural Languaes")
```

5. Use the **print()** function to print all tokens. Insert a new cell and add the following code to print the tokens:

```
print(sentence)
```

This code generates the following output:

```
['Ntural', 'Luanguage', 'Processin', 'deals', 'with', 'the', 'art',
'of', 'extracting', 'insightes', 'from', 'Natural', 'Languaes']
```

6. Now that we have got the tokens, loop through each token in **sentence**, correct the tokens, and assign them to a new variable. Insert a new cell and add the following code to implement this:

```
def correct_spelling(tokens):
    sentence_corrected = ' '.join([spell(word) \
                                   for word in tokens])
    return sentence_corrected
```

7. To print the correct sentence, insert a new cell and add the following code:

```
print(correct_spelling(sentence))
```

This code generates the following output:

```
['Natural', 'Language', 'Procession', 'deals', 'with', 'the', 'art',
 'of', 'extracting', 'insights', 'from', 'Natural', 'Languages']
```

In the preceding code snippet, we can see that most of the wrongly spelled words have been corrected. But the word "**Processin**" was wrongly converted into "**Procession**." It should have been "**Processing**." This happened because to change "**Processin**" to "**Procession**" or "**Processing**," an equal number of edits is required. To rectify this, we need to use other kinds of spelling correctors that are aware of context.

> **NOTE**
>
> To access the source code for this specific section, please refer to https://packt.live/38YVCKJ.
>
> You can also run this example online at https://packt.live/3gVpbj4.

In the next section, we will look at stemming, which is another form of text normalization.

STEMMING

In most languages, words get transformed into various forms when being used in a sentence. For example, the word "product" might get transformed into "production" when referring to the process of making something or transformed into "products" in plural form. It is necessary to convert these words into their base forms, as they carry the same meaning in any case. Stemming is the process that helps us to do so. If we look at the following figure, we get a perfect idea of how words get transformed into their base forms:

Figure 1.2: Stemming of the word "product"

To get a better understanding of stemming, let's perform a simple exercise.

In this exercise, we will be using two algorithms, called the porter stemmer and the snowball stemmer, provided by the NLTK library. The porter stemmer is a rule-based algorithm that transforms words to their base form by removing suffixes from words. The snowball stemmer is an improvement over the porter stemmer and is a little bit faster and uses less memory. In NLTK, this is done by the **stem()** method provided by the **PorterStemmer** class.

EXERCISE 1.07: USING STEMMING

In this exercise, we will pass a few words through the stemming process so that they get converted into their base forms. Follow these steps to implement this exercise:

1. Open a Jupyter Notebook.

2. Insert a new cell and add the following code to import the necessary libraries:

```
from nltk import stem
```

3. Now pass the following words as parameters to the **stem()** method. To implement this, insert a new cell and add the following code:

```
def get_stems(word,stemmer):
    return stemmer.stem(word)
porterStem = stem.PorterStemmer()

get_stems("production",porterStem)
```

4. When the input is **"production"**, the following output is generated:

```
'product'
```

5. Similarly, the following code would be used for the input **"coming"**.

```
get_stems("coming",porterStem)
```

We get the following output:

```
'come'
```

6. Similarly, the following code would be used for the input **"firing"**.

```
get_stems("firing",porterStem)
```

When the input is **"firing"**, the following output is generated:

```
'fire'
```

7. The following code would be used for the input **"battling"**.

```
get_stems("battling",porterStem)
```

If we give the input **"battling"**, the following output is generated:

```
'battl'
```

8. The following code will also generate the same output as above, for the input **"battling"**.

```
stemmer = stem.SnowballStemmer("english")
get_stems("battling",stemmer)
```

The output will be as follows:

```
'battl'
```

As you have seen while using the snowball stemmer, we have to provide the language as **"english"**. We can also use the stemmer for different languages such as Spanish, French, and many more. From the preceding code snippets, we can see that the entered words are converted into their base forms.

> **NOTE**
>
> To access the source code for this specific section, please refer to https://packt.live/2DLzisD.
>
> You can also run this example online at https://packt.live/30h147K.

In the next section, we will focus on **lemmatization**, which is another form of text normalization.

LEMMATIZATION

Sometimes, the stemming process leads to incorrect results. For example, in the last exercise, the word **battling** was transformed to **"battl"**, which is not a word. To overcome such problems with stemming, we make use of lemmatization. Lemmatization is the process of converting words to their base grammatical form, as in "battling" to "battle," rather than just randomly axing words. In this process, an additional check is made by looking through a dictionary to extract the base form of a word. Getting more accurate results requires some additional information; for example, PoS tags along with words will help in getting better results.

In the following exercise, we will be using **WordNetLemmatizer**, which is an NLTK interface of WordNet. WordNet is a freely available lexical English database that can be used to generate semantic relationships between words. NLTK's **WordNetLemmatizer** provides a method called **lemmatize()**, which returns the lemma (grammatical base form) of a given word using WordNet.

To put lemmatization into practice, let's perform an exercise where we'll use the **lemmatize()** function.

EXERCISE 1.08: EXTRACTING THE BASE WORD USING LEMMATIZATION

In this exercise, we will use the lemmatization process to produce the proper form of a given word. Follow these steps to implement this exercise:

1. Open a Jupyter Notebook.

2. Insert a new cell and add the following code to import the necessary libraries:

```
from nltk import download
download('wordnet')
from nltk.stem.wordnet import WordNetLemmatizer
```

3. Create an object of the **WordNetLemmatizer** class. Insert a new cell and add the following code to implement this:

```
lemmatizer = WordNetLemmatizer()
```

4. Bring the word to its proper form by using the **lemmatize()** method of the **WordNetLemmatizer** class. Insert a new cell and add the following code to implement this:

```
def get_lemma(word):
    return lemmatizer.lemmatize(word)

get_lemma('products')
```

With the input **products**, the following output is generated:

```
'product'
```

5. Similarly, use the input as **production** now:

```
get_lemma('production')
```

With the input **production**, the following output is generated:

```
'production'
```

6. Similarly, use the input as **coming** now:

```
get_lemma('coming')
```

With the input **coming**, the following output is generated:

```
'coming'
```

Hence, we have learned how to use the lemmatization process to transform a given word into its base form.

> **NOTE**
>
> To access the source code for this specific section, please refer to https://packt.live/3903ETS.
>
> You can also run this example online at https://packt.live/2Wlqu33.

In the next section, we will look at another preprocessing step in NLP: **named entity recognition (NER)**.

NAMED ENTITY RECOGNITION (NER)

NER is the process of extracting important entities, such as person names, place names, and organization names, from some given text. These are usually not present in dictionaries. So, we need to treat them differently. The main objective of this process is to identify the named entities (such as proper nouns) and map them to categories, which are already defined. For example, categories might include names of people, places, and so on.

NER has found use in many NLP tasks, including assigning tags to news articles, search algorithms, and more. NER can analyze a news article and extract the major people, organizations, and places discussed in it and assign them as tags for new articles.

In the case of search algorithms, let's suppose we have to create a search engine, meant specifically for books. If we were to submit a given query for all the words, the search would take a lot of time. Instead, if we extract the top entities from all the books using NER and run a search query on the entities rather than all the content, the speed of the system would increase dramatically.

To get a better understanding of this process, we'll perform an exercise. Before moving on to the exercise, let me introduce you to chunking, which we are going to use in the following exercise. Chunking is the process of grouping words together into chunks, which can be further used to find noun groups and verb groups, or can also be used for sentence partitioning.

EXERCISE 1.09: TREATING NAMED ENTITIES

In this exercise, we will find the named entities in a given sentence. Follow these steps to implement this exercise using the following sentence:

"We are reading a book published by Packt which is based out of Birmingham."

1. Open a Jupyter Notebook.

2. Insert a new cell and add the following code to import the necessary libraries:

```
from nltk import download
from nltk import pos_tag
from nltk import ne_chunk
from nltk import word_tokenize
download('maxent_ne_chunker')
download('words')
```

3. Declare the **sentence** variable and assign it a string. Insert a new cell and add the following code to implement this:

```
sentence = "We are reading a book published by Packt "\
           "which is based out of Birmingham."
```

4. To find the named entities from the preceding text, insert a new cell and add the following code:

```
def get_ner(text):
    i = ne_chunk(pos_tag(word_tokenize(text)), binary=True)
    return [a for a in i if len(a)==1]
get_ner(sentence)
```

This code generates the following output:

```
[Tree('NE', [('Packt', 'NNP')]), Tree('NE', [('Birmingham', 'NNP')])]
```

In the preceding code, we can see that the code identifies the named entities "**Packt**" and "**Birmingham**" and maps them to an already-defined category, "**NNP**."

> **NOTE**
>
> To access the source code for this specific section, please refer to https://packt.live/3ezeukC.
>
> You can also run this example online at https://packt.live/32rsOJs.

In the next section, we will focus on word sense disambiguation, which helps us to identify the right sense of any word.

WORD SENSE DISAMBIGUATION

There's a popular saying: "A man is known by the company he keeps." Similarly, a word's meaning depends on its association with other words in a sentence. This means two or more words with the same spelling may have different meanings in different contexts. This often leads to ambiguity. Word sense disambiguation is the process of mapping a word to the sense that it should carry. We need to disambiguate words based on the sense they carry so that they can be treated as different entities when being analyzed. The following figure displays a perfect example of how ambiguity is caused due to the usage of the same word in different sentences:

He knows how to **play** Harmonica.

We **play** only soccer.

Please **play** the next song.

What does **play** mean here?

Figure 1.3: Word sense disambiguation

One of the algorithms to solve word sense disambiguation is the Lesk algorithm. It has a huge corpus in the background (generally **WordNet** is used) that contains definitions of all the possible synonyms of all the possible words in a language. Then it takes a word and the context as input and finds a match between the context and all the definitions of the word. The meaning with the highest number of matches with the context of the word will be returned.

For example, suppose we have a sentence such as "We play only soccer" in a given text. Now, we need to find the meaning of the word "play" in this sentence. In the Lesk algorithm, each word with ambiguous meaning is saved in background synsets. In this case, the word "play" will be saved with all possible definitions. Let's say we have two definitions of the word "play":

1. Play: Participating in a sport or game

2. Play: Using a musical instrument

Then, we will find the similarity between the context of the word "play" in the text and both of the preceding definitions using text similarity techniques. The definition best suited to the context of "play" in the sentence will be considered the meaning or definition of the word. In this case, we will find that our first definition fits best in context, as the words "sport" and "game" are present in the preceding sentences.

In the next exercise, we will be using the Lesk module from NLTK. It takes a sentence and the word as input, and returns the meaning or definition of the word. The output of the Lesk method is **synset**, which contains the ID of the matched definition. These IDs can be matched with their definitions using the **definition()** method of **wsd.synset('word')**.

To get a better understanding of this process, let's look at an exercise.

EXERCISE 1.10: WORD SENSE DISAMBIGUATION

In this exercise, we will find the sense of the word "bank" in two different sentences. Follow these steps to implement this exercise:

1. Open a Jupyter Notebook.

2. Insert a new cell and add the following code to import the necessary libraries:

```
import nltk
nltk.download('wordnet')
from nltk.wsd import lesk
from nltk import word_tokenize
```

3. Declare two variables, **sentence1** and **sentence2**, and assign them with appropriate strings. Insert a new cell and the following code to implement this:

```
sentence1 = "Keep your savings in the bank"
sentence2 = "It's so risky to drive over the banks of the road"
```

4. To find the sense of the word "bank" in the preceding two sentences, use the Lesk algorithm provided by the **nltk.wsd** library. Insert a new cell and add the following code to implement this:

```
def get_synset(sentence, word):
    return lesk(word_tokenize(sentence), word)
get_synset(sentence1,'bank')
```

This code generates the following output:

```
Synset('savings_bank.n.02')
```

5. Here, **savings_bank.n.02** refers to a container for keeping money safely at home. To check the other sense of the word "bank," write the following code:

```
get_synset(sentence2,'bank')
```

This code generates the following output:

```
Synset('bank.v.07')
```

Here, **bank.v.07** refers to a slope in the turn of a road.

Thus, with the help of the Lesk algorithm, we were able to identify the sense of a word in whatever context.

> **NOTE**
>
> To access the source code for this specific section, please refer to https://packt.live/399JCq5.
>
> You can also run this example online at https://packt.live/30haCQ6.

In the next section, we will focus on **sentence boundary detection**, which helps detect the start and end points of sentences.

SENTENCE BOUNDARY DETECTION

Sentence boundary detection is the method of detecting where one sentence ends and another begins. If you are thinking that this sounds pretty easy, as a period (.) or a question mark (?) denotes the end of a sentence and the beginning of another sentence, then you are wrong. There can also be instances where the letters of acronyms are separated by full stops, for instance. Various analyses need to be performed at a sentence level; detecting the boundaries of sentences is essential.

An exercise will provide us with a better understanding of this process.

EXERCISE 1.11: SENTENCE BOUNDARY DETECTION

In this exercise, we will extract sentences from a paragraph. To do so, we'll be using the **sent_tokenize()** method, which is used to detect sentence boundaries. The following steps need to be performed:

1. Open a Jupyter Notebook.

2. Insert a new cell and add the following code to import the necessary libraries:

```
import nltk
from nltk.tokenize import sent_tokenize
```

3. Use the **sent_tokenize()** method to detect sentences in some given text. Insert a new cell and add the following code to implement this:

```
def get_sentences(text):
    return sent_tokenize(text)
get_sentences("We are reading a book. Do you know who is "\
              "the publisher? It is Packt. Packt is based "\
              "out of Birmingham.")
```

This code generates the following output:

```
['We are reading a book.'
 'Do you know who is the publisher?'
 'It is Packt.',
 'Packt is based out of Birmingham.']
```

4. Use the **sent_tokenize()** method for text that contains periods (.) other than those found at the ends of sentences:

```
get_sentences("Mr. Donald John Trump is the current "\
              "president of the USA. Before joining "\
              "politics, he was a businessman.")
```

The code will generate the following output:

```
['Mr. Donald John Trump is the current president of the USA.',
 'Before joining politics, he was a businessman.']
```

As you can see in the code, the `sent_tokenize` method is able to differentiate between the period (.) after "Mr" and the one used to end the sentence. We have covered all the preprocessing steps that are involved in NLP.

> **NOTE**
>
> To access the source code for this specific section, please refer to https://packt.live/2ZseU86.
>
> You can also run this example online at https://packt.live/2CC8Ukp.

Now, using the knowledge we've gained, let's perform an activity.

ACTIVITY 1.01: PREPROCESSING OF RAW TEXT

We have a text corpus that is in an improper format. In this activity, we will perform all the preprocessing steps that were discussed earlier to get some meaning out of the text.

> **NOTE**
>
> The text corpus, `file.txt`, can be found at this location: https://packt.live/30cu54z
>
> After downloading the file, place it in the same directory as the notebook.

Follow these steps to implement this activity:

1. Import the necessary libraries.

2. Load the text corpus to a variable.

3. Apply the tokenization process to the text corpus and print the first 20 tokens.

4. Apply spelling correction on each token and print the initial 20 corrected tokens as well as the corrected text corpus.

5. Apply PoS tags to each of the corrected tokens and print them.

6. Remove stop words from the corrected token list and print the initial 20 tokens.

7. Apply stemming and lemmatization to the corrected token list and then print the initial 20 tokens.

8. Detect the sentence boundaries in the given text corpus and print the total number of sentences.

> **NOTE**
>
> The solution to this activity can be found on page 366.

We have learned about and achieved the preprocessing of given data. By now, you should be familiar with what NLP is and what basic preprocessing steps are needed to carry out any NLP project. In the next section, we will focus on the different phases of an NLP project.

KICK STARTING AN NLP PROJECT

We can divide an NLP project into several sub-projects or phases. These phases are completed in a particular sequence. This tends to increase the overall efficiency of the process, as memory usage changes from one phase to the next. An NLP project has to go through six major phases, which are outlined in the following figure:

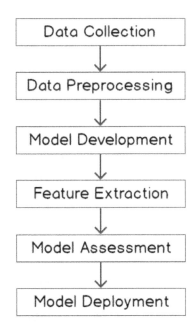

Figure 1.4: Phases of an NLP project

Suppose you are working on a project in which you need to classify emails as important and unimportant. We will explain how this is carried out by discussing each phase in detail.

DATA COLLECTION

This is the initial phase of any NLP project. Our sole purpose is to collect data as per our requirements. For this, we may either use existing data, collect data from various online repositories, or create our own dataset by crawling the web. In our case, we will collect different email data. We can even get this data from our personal emails as well, to start with.

DATA PREPROCESSING

Once the data is collected, we need to clean it. For the process of cleaning, we will make use of the different preprocessing steps that we have learned about in this chapter. It is necessary to clean the collected data to ensure effectiveness and accuracy. In our case, we will follow these preprocessing steps:

1. Converting all the text data to lowercase

2. Stop word removal

3. Text normalization, which will include replacing all numbers with some common term and replacing punctuation with empty strings

4. Stemming and lemmatization

FEATURE EXTRACTION

Computers understand only binary digits: 0 and 1. As such, every instruction we feed into a computer gets transformed into binary digits. Similarly, machine learning models tend to understand only numeric data. Therefore, it becomes necessary to convert text data into its equivalent numerical form.

To convert every email into its equivalent numerical form, we will create a dictionary of all the unique words in our data and assign a unique index to each word. Then, we will represent every email with a list having a length equal to the number of unique words in the data. The list will have 1 at the indices of words that are present in the email and 0 at the other indices. This is called one-hot encoding. We will learn more about this in coming chapters.

MODEL DEVELOPMENT

Once the feature set is ready, we need to develop a suitable model that can be trained to gain knowledge from the data. These models are generally statistical, machine learning-based, deep learning-based, or reinforcement learning-based. In our case, we will build a model that is capable of differentiating between important and unimportant emails.

MODEL ASSESSMENT

After developing a model, it is essential to benchmark it. This process of benchmarking is known as model assessment. In this step, we will evaluate the performance of our model by comparing it to others. This can be done by using different parameters or metrics. These parameters include precision, recall, and accuracy. In our case, we will evaluate the newly created model by seeing how well it performs at classifying emails as important and unimportant.

MODEL DEPLOYMENT

This is the final stage for most industrial NLP projects. In this stage, the models are put into production. They are either integrated into an existing system or new products are created by keeping this model as a base. In our case, we will deploy our model to production, so that it can classify emails as important and unimportant in real time.

SUMMARY

In this chapter, we learned about the basics of NLP and how it differs from text analytics. We covered the various preprocessing steps that are included in NLP, such as tokenization, PoS tagging, stemming, lemmatization, and more. We also looked at the different phases an NLP project has to pass through, from data collection to model deployment.

In the next chapter, you will learn about the different methods of extracting features from unstructured text, such as TF-IDF and bag of words. You will also learn about NLP tasks such as tokenization, lemmatization, and stemming in more detail. Furthermore, text visualization techniques such as word clouds will be introduced.

2

FEATURE EXTRACTION METHODS

OVERVIEW

In this chapter, you will be able to categorize data based on its content and structure. You will be able to describe preprocessing steps in detail and implement them to clean up text data. You will learn about feature engineering and calculate the similarity between texts. Once you understand these concepts, you will be able to use word clouds and some other techniques to visualize text.

INTRODUCTION

In the previous chapter, we learned about the concepts of **Natural Language Processing (NLP)** and text analytics. We also took a quick look at various preprocessing steps. In this chapter, we will learn how to make text understandable to machine learning algorithms.

As we know, to use a machine learning algorithm on textual data, we need a numerical or vector representation of text data since most of these algorithms are unable to work directly with plain text or strings. But before converting the text data into numerical form, we will need to pass it through some preprocessing steps such as tokenization, stemming, lemmatization, and stop-word removal.

So, in this chapter, we will learn a little bit more about these preprocessing steps and how to extract features from the preprocessed text and convert them into vectors. We will also explore two popular methods for feature extraction (Bag of Words and Term Frequency-Inverse Document Frequency), as well as various methods for finding similarity between different texts. By the end of this chapter, you will have gained an in-depth understanding of how text data can be visualized.

TYPES OF DATA

To deal with data effectively, we need to understand the various forms in which it exists. First, let's explore the types of data that exist. There are two main ways to categorize data (by structure and by content), as explained in the upcoming sections.

CATEGORIZING DATA BASED ON STRUCTURE

Data can be divided on the basis of structure into three categories, namely, structured, semi-structured, and unstructured data, as shown in the following diagram:

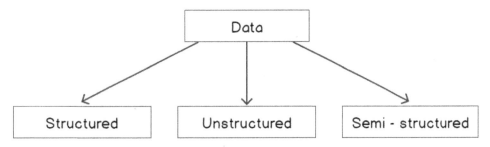

Figure 2.1: Categorization based on content

These three categories are as follows:

- **Structured data**: This is the most organized form of data. It is represented in tabular formats such as Excel files and **Comma-Separated Value** (**CSV**) files. The following image shows what structured data usually looks like:

Name	Age	Location
Ram	25	Delhi
Shyam	28	Banglore
Jon	35	Kolkata
Madhu	28	Mumbai
Hari	56	Chennai

Figure 2.2: Structured data

The preceding table contains information about five people, with each row representing a person and each column representing one of their attributes.

- **Semi-structured data**: This type of data is not presented in a tabular structure, but it can be transformed into a table. Here, information is usually stored between tags following a definite pattern. XML and HTML files can be referred to as semi-structured data. The following screenshot shows how semi-structured data can appear:

```
<student>
    <name>
        Jagat
    </name>
    <roll_number>
        3
    </roll_number>
    <rank>
        1
    </rank>
    <qualification>
        <qualification1>
            B.Tech
        </qualification1>
        <qualification2>
            M.Tech
        </qualification2>
    </qualification>
</student>
<student>
    <name>
        Jani
    </name>
    <roll_number>
        5
    </roll_number>
    <rank>
        3
    </rank>
    <qualification>
        <qualification1>
            B.A
        </qualification1>
    </qualification>
</student>
```

Figure 2.3: Semi-structured data

The format shown in the preceding screenshot is called markup language format. Here, the data is stored between tags, hierarchically. It is a universally accepted format, and there are a lot of parsers available that can convert this data into structured data.

- **Unstructured data**: This type of data is the most difficult to deal with. Machine learning algorithms would find it difficult to comprehend unstructured data without any loss of information. Text corpora and images are examples of unstructured data. The following image shows what unstructured data looks like:

We have three employees in Block A named James, Noah and Charlie. Their ages are 34,32 and 45 respectively. Charlie is from New Jersey while as Noah and James come from Waikiki.

Figure 2.4: Unstructured data

This is called unstructured data because if we want to get employee details from the preceding text snippet with our program, we will not be able to do so by simple parsing. We have to make our algorithm understand the semantics of the language to make it able to extract information from this.

CATEGORIZING DATA BASED ON CONTENT

Data can be divided into four categories based on content, as shown in the following diagram:

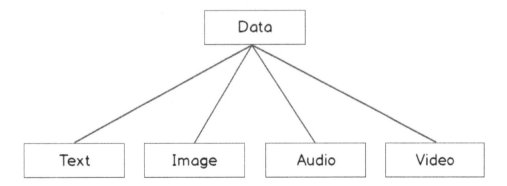

Figure 2.5: Categorizing data based on structure

Let's look at each category here:

- **Text data**: This refers to text corpora consisting of written sentences. This type of data can only be read. An example would be the text corpus of a book.

- **Image data**: This refers to pictures that are used to communicate messages. This type of data can only be seen.

- **Audio data**: This refers to voice recordings, music, and so on. This type of data can only be heard.

- **Video data**: A continuous series of images coupled with audio forms a video. This type of data can be seen as well as heard.

With that, we have learned about the different types of data and their categorization on the basis of structure and content. When dealing with unstructured data, it is necessary to clean it first. In the next section, we will look into some of the preprocessing steps for cleaning data.

CLEANING TEXT DATA

The text data that we are going to discuss here is unstructured text data, which consists of written sentences. Most of the time, this text data cannot be used as it is for analysis because it contains some noisy elements, that is, elements that do not really contribute much to the meaning of the sentence at all. These noisy elements need to be removed because they do not contribute to the meaning and semantics of the text. If they're not removed, they can not only waste system memory and processing time, but also negatively impact the accuracy of the results. Data cleaning is the art of extracting meaningful portions from data by eliminating unnecessary details. Consider the sentence, "He tweeted, *'Live coverage of General Elections available at this.tv/show/ge2019. _/_ Please tune in :) '.* "

In this example, to perform NLP tasks on the sentence, we will need to remove the emojis, punctuation, and stop words, and then change the words into their base grammatical form.

To achieve this, methods such as stopword removal, tokenization, and stemming are used. We will explore them in detail in the upcoming sections. Before we do so, let's get acquainted with some basic NLP libraries that we will be using here:

- **Re**: This is a standard Python library that's used for string searching and string manipulation. It contains methods such as **match()**, **search()**, **findall()**, **split()**, and **sub()**, which are used for basic string matching, searching, replacing, and more, using regular expressions. A regular expression is nothing but a set of characters in a specific order that represents a pattern. This pattern is searched for in the texts.

- **textblob**: This is an open source Python library that provides different methods for performing various NLP tasks such as tokenization and PoS tagging. It is similar to **nltk**, which was introduced in *Chapter 1, Introduction to Natural Language Processing*. It is built on the top of **nltk** and is much simpler as it has an easier to use interface and excellent documentation. In projects that don't involve a lot of complexity, it should be preferable to **nltk**.

- **keras**: This is an open source, high-level neural network library that's was developed on top of another neural network library called **TensorFlow**. In addition to neural network functionality, it also provides methods for basic text processing and NLP tasks.

TOKENIZATION

Tokenization and word tokenizers were briefly described in *Chapter 1, Introduction to Natural Language Processing*. Tokenization is the process of splitting sentences into their constituents; that is, words and punctuation. Let's perform a simple exercise to see how this can be done using various packages.

EXERCISE 2.01: TEXT CLEANING AND TOKENIZATION

In this exercise, we will clean some text and extract the tokens from it. Follow these steps to complete this exercise:

1. Open a Jupyter Notebook.

2. Import the **re** package:

```
import re
```

3. Create a method called **clean_text()** that will delete all characters other than digits, alphabetical characters, and whitespaces from the text and split the text into tokens. For this, we will use the text which matches with all non-alphanumeric characters, and we will replace all of them with an empty string:

```
def clean_text(sentence):
    return re.sub(r'([^\s\w]|_)+', ' ', sentence).split()
```

4. Store the sentence to be cleaned in a variable named **sentence** and pass it through the preceding function. Add the following code to this: implement

```
sentence = 'Sunil tweeted, "Witnessing 70th Republic Day "\
          "of India from Rajpath, New Delhi. "\
          "Mesmerizing performance by Indian Army! "\
          "Awesome airshow! @india_official "\
          "@indian_army #India #70thRepublic_Day. "\
          "For more photos ping me sunil@photoking.com :)"'
clean_text(sentence)
```

The preceding command fragments the string wherever any blank space is present. The output should be as follows:

```
['Sunil',
 'tweeted',
 'Witnessing',
 '70th',
 'Republic',
 'Day',
 'of',
 'India',
 'from',
 'Rajpath',
 'New',
 'Delhi',
 'Mesmerizing',
 'performance',
 'by',
 'Indian',
 'Army',
 'Awesome',
 'airshow',
 'india',
 'official',
 'indian',
 'army',
 'India',
 '70thRepublic',
 'Day',
 'For',
 'more',
 'photos',
 'ping',
 'me',
 'sunil',
 'photoking',
 'com']
```

Figure 2.6: Fragmented string

With that, we have learned how to extract tokens from text. Often, extracting each token separately does not help. For instance, consider the sentence, "I don't hate you, but your behavior." Here, if we process each of the tokens, such as "hate" and "behavior," separately, then the true meaning of the sentence would not be comprehended. In this case, the context in which these tokens are present becomes essential. Thus, we consider **n** consecutive tokens at a time. **n-grams** refers to the grouping of **n** consecutive tokens together.

> **NOTE**
>
> To access the source code for this specific section, please refer to https://packt.live/2CQikt7.
>
> You can also run this example online at https://packt.live/33cn0nF.

Next, we will look at an exercise where n-grams can be extracted from a given text.

EXERCISE 2.02: EXTRACTING N-GRAMS

In this exercise, we will extract n-grams using three different methods. First, we will use custom-defined functions, and then the **nltk** and **textblob** libraries. Follow these steps to complete this exercise:

1. Open a Jupyter Notebook.

2. Import the **re** package and create a custom-defined function, which we can use to extract **n**-grams. Add the following code to do this:

```
import re
def n_gram_extractor(sentence, n):
    tokens = re.sub(r'([^\s\w]|_)+', ' ', sentence).split()
    for i in range(len(tokens)-n+1):
        print(tokens[i:i+n])
```

In the preceding function, we are splitting the sentence into tokens using regex, then looping over the tokens, taking **n** consecutive tokens at a time.

3. If **n** is 2, two consecutive tokens will be taken, resulting in bigrams. To check the bigrams, we pass the function the text and with **n=2**. Add the following code to do this:

```
n_gram_extractor('The cute little boy is playing with the kitten.', \
                 2)
```

The preceding code generates the following output:

```
['The', 'cute']
['cute', 'little']
['little', 'boy']
['boy', 'is']
['is', 'playing']
['playing', 'with']
['with', 'the']
['the', 'kitten']
```

4. To check the trigrams, we pass the function with the text and with **n=3**. Add the following code to do this:

```
n_gram_extractor('The cute little boy is playing with the kitten.', \
                 3)
```

The preceding code generates the following output:

```
['The', 'cute', 'little']
['cute', 'little', 'boy']
['little', 'boy', 'is']
['boy', 'is', 'playing']
['is', 'playing', 'with']
['playing', 'with', 'the']
['with', 'the', 'kitten']
```

5. To check the bigrams using the **nltk** library, add the following code:

```
from nltk import ngrams
list(ngrams('The cute little boy is playing with the kitten.'\
            .split(), 2))
```

The preceding code generates the following output:

```
[('The', 'cute'),
 ('cute', 'little'),
 ('little', 'boy'),
 ('boy', 'is'),
 ('is', 'playing'),
 ('playing', 'with'),
 ('with', 'the'),
 ('the', 'kitten')]
```

6. To check the trigrams using the **nltk** library, add the following code:

```
list(ngrams('The cute little boy is playing with the
kitten.'.split(), 3))
```

The preceding code generates the following output:

```
[('The', 'cute', 'little'),
 ('cute', 'little', 'boy'),
 ('little', 'boy', 'is'),
 ('boy', 'is', 'playing'),
 ('playing', 'with', 'the'),
 ('with', 'the', 'kitten.')]
```

7. To check the bigrams using the **textblob** library, add the following code:

```
!pip install -U textblob
from textblob import TextBlob
blob = TextBlob("The cute little boy is playing with the kitten.")
blob.ngrams(n=2)
```

The preceding code generates the following output:

```
[WordList(['The', 'cute']),
 WordList(['cute', 'little']),
 WordList(['little', 'boy']),
 WordList(['boy', 'is']),
 WordList(['is', 'playing']),
 WordList(['playing', 'with']),
 WordList(['with', 'the']),
 WordList(['the', 'kitten'])]
```

8. To check the trigrams using the **textblob** library, add the following code:

```
blob.ngrams(n=3)
```

The preceding code generates the following output:

```
[WordList(['The', 'cute', 'little']),
 WordList(['cute', 'little', 'boy']),
 WordList(['little', 'boy', 'is']),
 WordList(['boy', 'is' 'playing']),
 WordList(['is', 'playing' 'with']),
 WordList(['playing', 'with' 'the']),
 WordList(['with', 'the' 'kitten'])]
```

In this exercise, we learned how to generate n-grams using various methods.

> **NOTE**
>
> To access the source code for this specific section, please refer to https://packt.live/2PabHUK.
>
> You can also run this example online at https://packt.live/2XbjFRX.

EXERCISE 2.03: TOKENIZING TEXT WITH KERAS AND TEXTBLOB

In this exercise, we will use **keras** and **textblob** to tokenize texts. Follow these steps to complete this exercise:

1. Open a Jupyter Notebook and insert a new cell.

2. Import the **keras** and **textblob** libraries and declare a variable named **sentence**, as follows.

```
from keras.preprocessing.text import text_to_word_sequence
from textblob import TextBlob

sentence = 'Sunil tweeted, "Witnessing 70th Republic Day "\
           "of India from Rajpath, New Delhi. "\
           "Mesmerizing performance by Indian Army! "\
           "Awesome airshow! @india_official "\
           "@indian_army #India #70thRepublic_Day. "\
           "For more photos ping me sunil@photoking.com :)"'
```

3. To tokenize using the **keras** library, add the following code:

```
def get_keras_tokens(text):
    return text_to_word_sequence(text)


get_keras_tokens(sentence)
```

The preceding code generates the following output:

```
['sunil',
 'tweeted',
 'witnessing',
 '70th',
 'republic',
 'day',
 'of',
 'india',
 'from',
 'rajpath',
 'new',
 'delhi',
 'mesmerizing',
 'performancesby',
 'indian',
 'army',
 'awesome',
 'airshow',
 'india',
 'official',
 'indian',
 'army',
 'india',
 '70threpublic',
 'day',
 'for',
 'more',
 'photos',
 'ping',
 'me',
 'sunil',
 'photoking',
 'com']
```

Figure 2.7: Tokenization using Keras

4. To tokenize using the **textblob** library, add the following code:

```
def get_textblob_tokens(text):
    blob = TextBlob(text)
    return blob.words


get_textblob_tokens(sentence)
```

The preceding code generates the following output:

```
WordList(['Sunil', 'tweeted', 'Witnessing', '70th', 'Republic', 'Day', 'of', 'India', 'from', 'Rajpa
th', 'New', 'Delhi', 'Mesmerizing', 'performancesby', 'Indian', 'Army', 'Awesome', 'airshow', 'india
_official', 'indian_army', 'India', '70thRepublic_Day', 'For', 'more', 'photos', 'ping', 'me', 'suni
l', 'photoking.com'])
```

Figure 2.8: Tokenization using textblob

With that, we have learned how to tokenize texts using the **keras** and **textblob** libraries.

> **NOTE**
>
> To access the source code for this specific section, please refer to https://packt.live/3393hFi.
>
> You can also run this example online at https://packt.live/39Dtu09.

In the next section, we will discuss the different types of tokenizers.

TYPES OF TOKENIZERS

There are different types of tokenizers that come in handy for specific tasks. Let's look at the ones provided by **nltk** one by one:

- **Whitespace tokenizer**: This is the simplest type of tokenizer. It splits a string wherever a space, tab, or newline character is present.

- **Tweet tokenizer**: This is specifically designed for tokenizing tweets. It takes care of all the special characters and emojis used in tweets and returns clean tokens.

- **MWE tokenizer**: MWE stands for Multi-Word Expression. Here, certain groups of multiple words are treated as one entity during tokenization, such as "United States of America," "People's Republic of China," "not only," and "but also." These predefined groups are added at the beginning with **mwe()** methods.

- **Regular expression tokenizer**: These tokenizers are developed using regular expressions. Sentences are split based on the occurrence of a specific pattern (a regular expression).

- **WordPunctTokenizer**: This splits a piece of text into a list of alphabetical and non-alphabetical characters. It actually splits text into tokens using a fixed **regex**, that is, `'\w+|[^\w\s]+'`.

Now that we have learned about the different types of tokenizers, in the next section, we will carry out an exercise to get a better understanding of them.

EXERCISE 2.04: TOKENIZING TEXT USING VARIOUS TOKENIZERS

In this exercise, we will use different tokenizers to tokenize text. Perform the following steps to implement this exercise:

1. Open a Jupyter Notebook.

2. Insert a new cell and the following code to import all the tokenizers and declare a variable sentence:

```
from nltk.tokenize import TweetTokenizer
from nltk.tokenize import MWETokenizer
from nltk.tokenize import RegexpTokenizer
from nltk.tokenize import WhitespaceTokenizer
from nltk.tokenize import WordPunctTokenizer

sentence = 'Sunil tweeted, "Witnessing 70th Republic Day "\
           "of India from Rajpath, New Delhi. "\
           "Mesmerizing performance by Indian Army! "\
           "Awesome airshow! @india_official "\
           "@indian_army #India #70thRepublic_Day. "\
           "For more photos ping me sunil@photoking.com :)"'
```

3. To tokenize the text using **TweetTokenizer**, add the following code:

```
def tokenize_with_tweet_tokenizer(text):
    # Here will create an object of tweetTokenizer
    tweet_tokenizer = TweetTokenizer()
    """
    Then we will call the tokenize method of
    tweetTokenizer which will return token list of sentences.
    """
    return tweet_tokenizer.tokenize(text)
tokenize_with_tweet_tokenizer(sentence)
```

> **NOTE**
>
> The # symbol in the code snippet above denotes a code comment.
> Comments are added into code to help explain specific bits of logic.

The preceding code generates the following output:

```
['Sunil',
 'tweeted',
 ',',
 '"',
 'Witnessing',
 '70th',
 'Republic',
 'Day',
 'of',
 'India',
 'from',
 'Rajpath',
 ',',
 'New',
 'Delhi',
 '.',
 'Mesmerizing',
 'performance',
 'by',
 'Indian',
 'Army',
 '!',
 'Awesome',
 'airshow',
 '!',
 '@india_official',
 '@indian_army',
 '#India',
 '#70thRepublic_Day',
 '.',
 'For',
 'more',
 'photos',
 'ping',
 'me',
 'sunil@photoking.com',
 ':)',
 '"']
```

Figure 2.9: Tokenization using TweetTokenizer

As you can see, the hashtags, emojis, websites, and Twitter IDs are extracted as single tokens. If we had used the white space tokenizer, we would have got hash, dots, and the @ symbol as separate tokens.

4. To tokenize the text using **MWETokenizer**, add the following code:

```
def tokenize_with_mwe(text):
    mwe_tokenizer = MWETokenizer([('Republic', 'Day')])
    mwe_tokenizer.add_mwe(('Indian', 'Army'))
    return mwe_tokenizer.tokenize(text.split())


tokenize_with_mwe(sentence)
```

The preceding code generates the following output:

```
['Sunil',
 'tweeted,',
 '"Witnessing',
 '70th',
 'Republic_Day',
 'of',
 'India',
 'from',
 'Rajpath,',
 'New',
 'Delhi.',
 'Mesmerizing',
 'performance',
 'by',
 'Indian',
 'Army!',
 'Awesome',
 'airshow!',
 '@india_official',
 '@indian_army',
 '#India',
 '#70thRepublic_Day.',
 'For',
 'more',
 'photos',
 'ping',
 'me',
 'sunil@photoking.com',
 ':)"']
```

Figure 2.10: Tokenization using the MWE tokenizer

In the preceding screenshot, the words "Indian" and "Army!", which should have been treated as a single identity, were treated separately. This is because "Army!" (not "Army") is treated as a token. Let's see how this can be fixed in the next step.

5. Add the following code to fix the issues in the previous step:

```
tokenize_with_mwe(sentence.replace('!',''))
```

The preceding code generates the following output:

```
['Sunil',
 'tweeted,',
 '"Witnessing',
 '70th',
 'Republic_Day',
 'of',
 'India',
 'from',
 'Rajpath,',
 'New',
 'Delhi.',
 'Mesmerizing',
 'performance',
 'by',
 'Indian_Army',
 'Awesome',
 'airshow',
 '@india_official',
 '@indian_army',
 '#India',
 '#70thRepublic_Day.',
 'For',
 'more',
 'photos',
 'ping',
 'me',
 'sunil@photoking.com',
 ':)"']
```

Figure 2.11: Tokenization using the MWE tokenizer after removing the "!" sign

Here, we can see that instead of being treated as separate tokens, "Indian" and "Army" are treated as a single entity.

6. To tokenize the text using the regular expression tokenizer, add the following code:

```
def tokenize_with_regex_tokenizer(text):
    reg_tokenizer = RegexpTokenizer('\w+|\$[\d\.]+|\S+')
    return reg_tokenizer.tokenize(text)


tokenize_with_regex_tokenizer(sentence)
```

The preceding code generates the following output:

```
['Sunil',
 'tweeted',
 ',',
 '"Witnessing',
 '70th',
 'Republic',
 'Day',
 'of',
 'India',
 'from',
 'Rajpath',
 ',',
 'New',
 'Delhi',
 '.',
 'Mesmerizing',
 'performance',
 'by',
 'Indian',
 'Army',
 '!',
 'Awesome',
 'airshow',
 '!',
 '@india_official',
 '@indian_army',
 '#India',
 '#70thRepublic_Day.',
 'For',
 'more',
 'photos',
 'ping',
 'me',
 'sunil',
 '@photoking.com',
 ':)"']
```

Figure 2.12: Tokenization using the regular expression tokenizer

7. To tokenize the text using the whitespace tokenizer, add the following code:

```
def tokenize_with_wst(text):
    wh_tokenizer = WhitespaceTokenizer()
    return wh_tokenizer.tokenize(text)


tokenize_with_wst(sentence)
```

The preceding code generates the following output:

```
['Sunil',
 'tweeted,',
 '"Witnessing',
 '70th',
 'Republic',
 'Day',
 'of',
 'India',
 'from',
 'Rajpath,',
 'New',
 'Delhi.',
 'Mesmerizing',
 'performance',
 'by',
 'Indian',
 'Army!',
 'Awesome',
 'airshow!',
 '@india_official',
 '@indian_army',
 '#India',
 '#70thRepublic_Day.',
 'For',
 'more',
 'photos',
 'ping',
 'me',
 'sunil@photoking.com',
 ':)"']
```

Figure 2.13: Tokenization using the whitespace tokenizer

8. To tokenize the text using the Word Punct tokenizer, add the following code:

```
def tokenize_with_wordpunct_tokenizer(text):
    wp_tokenizer = WordPunctTokenizer()
    return wp_tokenizer.tokenize(text)


tokenize_with_wordpunct_tokenizer(sentence)
```

The preceding code generates the following output:

```
['Sunil',
 'tweeted',
 ',',
 '"',
 'Witnessing',
 '70th',
 'Republic',
 'Day',
 'of',
 'India',
 'from',
 'Rajpath',
 ',',
 'New',
 'Delhi',
 '.',
 'Mesmerizing',
 'performance',
 'by',
 'Indian',
 'Army',
 '!',
 'Awesome',
 'airshow',
 '!',
 '@',
 'india_official',
 '@',
 'indian_army',
 '#',
 'India',
 '#',
 '70thRepublic_Day',
 '.',
 'For',
 'more',
 'photos',
 'ping',
 'me',
 'sunil',
 '@',
 'photoking',
 '.',
 'com',
 ':)"']
```

Figure 2.14: Tokenization using the Word Punct tokenizer

In this section, we have learned about different tokenization techniques and their **nltk** implementation.

> **NOTE**
>
> To access the source code for this specific section, please refer to https://packt.live/3hSbDWi.
>
> You can also run this example online at https://packt.live/3hQi7oR.

Now, we're ready to use them in our programs.

STEMMING

In many languages, the base forms of words change when they're used in sentences. For example, the word "produce" can be written as "production" or "produced" or even "producing," depending on the context. The process of converting a word back into its base form is known as stemming. It is essential to do this, because without it, algorithms would treat two or more different forms of the same word as different entities, despite them having the same semantic meaning. So, the words "producing" and "produced" would be treated as different entities, which can lead to erroneous inferences. In Python, **RegexpStemmer** and **PorterStemmer** are the most widely used stemmers. Let's explore them one at a time.

REGEXPSTEMMER

RegexpStemmer uses regular expressions to check whether morphological or structural prefixes or suffixes are present. For instance, in many cases, verbs in the present continuous tense (the present tense form ending with "ing") can be restored to their base form simply by removing "ing" from the end; for example, "playing" becomes "play".

Let's complete the following exercise to get some hands-on experience with **RegexpStemmer**.

EXERCISE 2.05: CONVERTING WORDS IN THE PRESENT CONTINUOUS TENSE INTO BASE WORDS WITH REGEXPSTEMMER

In this exercise, we will use **RegexpStemmer** on text to convert words into their basic form by removing some generic suffixes such as "ing" and "ed". To use **nltk**'s **regex_stemmer**, we have to create an object of **RegexpStemmer** by passing the regex of the suffix or prefix and an integer, **min**, which indicates the minimum length of the stemmed string. Follow these steps to complete this exercise:

1. Open a Jupyter Notebook.

2. Insert a new cell and import **RegexpStemmer**:

```
from nltk.stem import RegexpStemmer
```

3. Use **regex_stemmer** to stem each word of the **sentence** variable. Add the following code to do this:

```
def get_stems(text):
    """
    Creating an object of RegexpStemmer, any string ending
    with the given regex 'ing$' will be removed.
    """

    regex_stemmer = RegexpStemmer('ing$', min=4)

    """
    The below code line will convert every word into its
    stem using regex stemmer and then join them with space.
    """
    return ' '.join([regex_stemmer.stem(wd) for \
                    wd in text.split()])

sentence = "I love playing football"
get_stems(sentence)
```

The preceding code generates the following output:

```
'I love play football'
```

As we can see, the word **playing** has been changed into its base form, **play**. In this exercise, we learned how we can perform stemming using **nltk**'s **RegexpStemmer**.

> **NOTE**
>
> To access the source code for this specific section, please refer to https://packt.live/3hRYUm6.
>
> You can also run this example online at https://packt.live/2D0Ztvk.

THE PORTER STEMMER

The Porter stemmer is the most common stemmer for dealing with English words. It removes various morphological and inflectional endings (such as suffixes, prefixes, and the plural "s") from English words. In doing so, it helps us extract the base form of a word from its variations. To get a better understanding of this, let's carry out a simple exercise.

EXERCISE 2.06: USING THE PORTER STEMMER

In this exercise, we will apply the Porter stemmer to some text. Follow these steps to complete this exercise:

1. Open a Jupyter Notebook.

2. Import **nltk** and any related packages and declare a **sentence** variable. Add the following code to do this:

```
from nltk.stem.porter import *

sentence = "Before eating, it would be nice to "\
           "sanitize your hands with a sanitizer"
```

3. Now, we'll make use of the Porter stemmer to stem each word of the **sentence** variables:

```
def get_stems(text):
    ps_stemmer = PorterStemmer()
    return ' '.join([ps_stemmer.stem(wd) for \
                    wd in text.split()])

get_stems(sentence)
```

The preceding code generates the following output:

```
'befor eating, it would be nice to sanit your hand wash with a sanit'
```

> **NOTE**
>
> To access the source code for this specific section, please refer to https://packt.live/2CUqelc.
>
> You can also run this example online at https://packt.live/2X8WUhD.

PorterStemmer is a generic rule-based stemmer that tries to convert a word into its basic form by removing common suffixes and prefixes of the English language.

Though stemming is a useful technique in NLP, it has a severe drawback. As we can see from this exercise, we find that, while **eating** has been converted into **eat** (which is its proper grammatical base form), the word **sanitize** has been converted into **sanit** (which isn't the proper grammatical base form). This may lead to some problems if we use it. To overcome this issue, there is another technique we can use called lemmatization.

LEMMATIZATION

As we saw in the previous section, there is a problem with stemming. It often generates meaningless words. Lemmatization deals with such cases by using vocabulary and analyzing the words' morphologies. It returns the base forms of words that can be found in dictionaries. Let's walk through a simple exercise to understand this better.

EXERCISE 2.07: PERFORMING LEMMATIZATION

In this exercise, we will perform lemmatization on some text. Follow these steps to complete this exercise:

1. Open a Jupyter Notebook.

2. Import **nltk** and its related packages, and then declare a **sentence** variable. Add the following code to implement this:

```
import nltk
from nltk.stem import WordNetLemmatizer
from nltk import word_tokenize
nltk.download('wordnet')
nltk.download('punkt')


sentence = "The products produced by the process today are "\
           "far better than what it produces generally."
```

3. To lemmatize the tokens, we extracted from the sentence, add the following code:

```
lemmatizer = WordNetLemmatizer()
def get_lemmas(text):
    lemmatizer = WordNetLemmatizer()
    return ' '.join([lemmatizer.lemmatize(word) for \
                     word in word_tokenize(text)])


get_lemmas(sentence)
```

The preceding code generates the following output:

```
'The product produced by the process today are far better than what
it produce generally.'
```

With that, we learned how to generate the lemma of a word. The lemma is the correct grammatical base form. They use the vocabulary to match the word to its correct nearest grammatical form.

> **NOTE**
>
> To access the source code for this specific section, please refer to https://packt.live/2X5JEKA.
>
> You can also run this example online at https://packt.live/30Zqt6v.

In the next section, we will deal with other kinds of word variations by looking at singularizing and pluralizing words using **textblob**.

EXERCISE 2.08: SINGULARIZING AND PLURALIZING WORDS

In this exercise, we will make use of the **textblob** library to singularize and pluralize words in the given text. Follow these steps to complete this exercise:

1. Open a Jupyter Notebook.

2. Import **TextBlob** and declare a **sentence** variable. Add the following code to implement this:

```
from textblob import TextBlob
sentence = TextBlob('She sells seashells on the seashore')
```

To check the list of words in the sentence, type the following code:

```
sentence.words
```

The preceding code generates the following output:

```
WordList(['She', 'sells', 'seashells', 'on', 'the', 'seashore'])
```

3. To singularize the third word in the sentence, type the following code:

```
def singularize(word):
    return word.singularize()


singularize(sentence.words[2])
```

The preceding code generates the following output:

```
'seashell'
```

4. To pluralize the fifth word in the given sentence, type the following code:

```
def pluralize(word):
    return word.pluralize()
pluralize(sentence.words[5])
```

The preceding code generates the following output:

```
'seashores'
```

> **NOTE**
>
> To access the source code for this specific section, please refer to https://packt.live/3gooUoQ.
>
> You can also run this example online at https://packt.live/309Gqrm.

Now, in the next section, we will learn about another preprocessing task: language translation.

LANGUAGE TRANSLATION

You might have used Google Translate before, which gives the exact translation of a word in another language; this is an example of language translation or machine translation. In Python, we can use **TextBlob** to translate text from one language into another. **TextBlob** provides a method called **translate()**, in which you have to pass text in the source language. The method will return the translated word in the destination language. Let's look at how this is done.

EXERCISE 2.09: LANGUAGE TRANSLATION

In this exercise, we will make use of the **TextBlob** library to translate a sentence from Spanish into English. Follow these steps to implement this exercise:

1. Open a Jupyter Notebook.

2. Import **TextBlob**, as follows:

```
from textblob import TextBlob
```

3. Make use of the **translate()** function of **TextBlob** to translate the input text from Spanish to English. Add the following code to do this:

```
def translate(text,from_l,to_l):
    en_blob = TextBlob(text)
    return en_blob.translate(from_lang=from_l, to=to_l)

translate(text='muy bien',from_l='es',to_l='en')
```

The preceding code generates the following output:

```
TextBlob("very well")
```

With that, we have seen how we can use **TextBlob** to translate from one language to another.

> **NOTE**
>
> To access the source code for this specific section, please refer to https://packt.live/2XquGiH.
>
> You can also run this example online at https://packt.live/3hQiVK8.

In the next section, we will look at another preprocessing task: stop-word removal.

STOP-WORD REMOVAL

Stop words, such as "am," "the," and "are," occur frequently in text data. Although they help us construct sentences properly, we can find the meaning even if we remove them. This means that the meaning of text can be inferred even without them. So, removing stop words from text is one of the preprocessing steps in NLP tasks. In Python, **nltk**, and **textblob**, text can be used to remove stop words from text. To get a better understanding of this, let's look at an exercise.

EXERCISE 2.10: REMOVING STOP WORDS FROM TEXT

In this exercise, we will remove the stop words from a given text. Follow these steps to complete this exercise:

1. Open a Jupyter Notebook.

2. Import **nltk** and declare a **sentence** variable with the text in question:

```
from nltk import word_tokenize
sentence = "She sells seashells on the seashore"
```

3. Define a **remove_stop_words** method and remove the custom list of stop words from the sentence by using the following lines of code:

```
def remove_stop_words(text,stop_word_list):
    return ' '.join([word for word in word_tokenize(text) \
                    if word.lower() not in stop_word_list])

custom_stop_word_list = ['she', 'on', 'the', 'am', 'is', 'not']
remove_stop_words(sentence,custom_stop_word_list)
```

The preceding code generates the following output:

```
'sells seashells seashore'
```

Thus, we've seen how stop words can be removed from a sentence.

> **NOTE**
>
> To access the source code for this specific section, please refer to https://packt.live/337aMwH.
>
> You can also run this example online at https://packt.live/30buvJF.

In the next activity, we'll put our knowledge of preprocessing steps into practice.

ACTIVITY 2.01: EXTRACTING TOP KEYWORDS FROM THE NEWS ARTICLE

In this activity, you will extract the most frequently occurring keywords from a sample news article.

> **NOTE**
>
> The new article that's being used for this activity can be found at https://packt.live/314mg1r.

The following steps will help you implement this activity:

1. Open a Jupyter Notebook.

2. Import **nltk** and any other necessary libraries.

3. Define some functions to help you load the text file, convert the string into lowercase, tokenize the text, remove the stop words, and perform stemming on all the remaining tokens. Finally, define a function to calculate the frequency of all these words.

4. Load **news_article.txt** using a Python file reader into a single string.

5. Convert the text string into lowercase.

6. Split the string into tokens using a white space tokenizer.

7. Remove any stop words.

8. Perform stemming on all the tokens.

9. Calculate the frequency of all the words after stemming.

> **NOTE**
>
> The solution to this activity can be found on page 373.

With that, we have learned about the various ways we can clean unstructured data. Now, let's examine the concept of extracting features from texts.

FEATURE EXTRACTION FROM TEXTS

As we already know, machine learning algorithms do not understand textual data directly. We need to represent the text data in numerical form or vectors. To convert each textual sentence into a vector, we need to represent it as a set of features. This set of features should uniquely represent the text, though, individually, some of the features may be common across many textual sentences. Features can be classified into two different categories:

- **General features**: These features are statistical calculations and do not depend on the content of the text. Some examples of general features could be the number of tokens in the text, the number of characters in the text, and so on.

- **Specific features**: These features are dependent on the inherent meaning of the text and represent the semantics of the text. For example, the frequency of unique words in the text is a specific feature.

Let's explore these in detail.

EXTRACTING GENERAL FEATURES FROM RAW TEXT

As we've already learned, general features refer to those that are not directly dependent on the individual tokens constituting a text corpus. Let's consider these two sentences: "The sky is blue" and "The pillar is yellow". Here, the sentences have the same number of words (a general feature)—that is, four. But the individual constituent tokens are different. Let's complete an exercise to understand this better.

EXERCISE 2.11: EXTRACTING GENERAL FEATURES FROM RAW TEXT

In this exercise, we will extract general features from input text. These general features include detecting the number of words, the presence of "wh" words (words beginning with "wh", such as "what" and "why") and the language in which the text is written. Follow these steps to implement this exercise:

1. Open a Jupyter Notebook.

2. Import the **pandas** library and create a DataFrame with four sentences. Add the following code to implement this:

```
import pandas as pd
from textblob import TextBlob



df = pd.DataFrame([['The interim budget for 2019 will '\
            'be announced on 1st February.'], \
            ['Do you know how much expectation '\
            'the middle-class working population '\
            'is having from this budget?'], \
            ['February is the shortest month '\
            'in a year.'], \
            ['This financial year will end on '\
            '31st March.']])
df.columns = ['text']
df.head()
```

The preceding code generates the following output:

	text
0	The interim budget for 2019 will be announced ...
1	Do you know how much expectation the middle-cl...
2	February is the shortest month in a year.
3	This financial year will end on 31st March.

Figure 2.15: DataFrame consisting of four sentences

3. Use the **apply()** function to iterate through each row of the column text, convert them into **TextBlob** objects, and extract words from them. Add the following code to implement this:

```
def add_num_words(df):
    df['number_of_words'] = df['text'].apply(lambda x : \
                           len(TextBlob(str(x)).words))
    return df
add_num_words(df)['number_of_words']
```

The preceding code generates the following output:

```
0      11
1      15
2       8
3       8
Name:  number_of_words, dtype: int64
```

The preceding code line will print the **number_of_words** column of the DataFrame to represent the number of words in each row.

4. Use the **apply()** function to iterate through each row of the column text, convert the text into **TextBlob** objects, and extract the words from them to check whether any of them belong to the list of "wh" words that has been declared. Add the following code to do so:

```
def is_present(wh_words, df):

    """
    The below line of code will find the intersection
    between set of tokens of every sentence and the
    wh_words and will return true if the length of
    intersection set is non-zero.
    """
    df['is_wh_words_present'] = df['text'].apply(lambda x : \
                          True if \
                          len(set(TextBlob(str(x)).\
                          words).intersection(wh_words))\
                          >0 else False)
    return df
```

```
wh_words = set(['why', 'who', 'which', 'what', \
                'where', 'when', 'how'])

is_present(wh_words, df)['is_wh_words_present']
```

The preceding code generates the following output:

```
0       False
1       True
2       False
3       False
Name:  is_wh_words_present, dtype: bool
```

The preceding code line will print the **is_wh_words_present** column that was added by the **is_present** method to **df**, which means for every row, we will see whether **wh_word** is present.

5. Use the **apply()** function to iterate through each row of the column text, convert them into **TextBlob** objects, and detect their languages:

```
def get_language(df):
    df['language'] = df['text'].apply(lambda x : \
                    TextBlob(str(x)).detect_language())
    return df
get_language(df)['language']
```

The preceding code generates the following output:

```
0       en
1       en
2       en
3       en
Name:  language, dtype: object
```

With that, we have learned how to extract general features from text data.

> **NOTE**
>
> To access the source code for this specific section, please refer to https://packt.live/2X9jLcS.
>
> You can also run this example online at https://packt.live/3fgrYSK.

Let's perform another exercise to get a better understanding of this.

EXERCISE 2.12: EXTRACTING GENERAL FEATURES FROM TEXT

In this exercise, we will extract various general features from documents. The dataset that we will be using here consists of random statements. Our objective is to find the frequency of various general features such as punctuation, uppercase and lowercase words, letters, digits, words, and whitespaces.

> **NOTE**
>
> The dataset that is being used in this exercise can be found at this link: https://packt.live/3k0qCPR.

1. Open a Jupyter Notebook.

2. Insert a new cell and add the following code to import the necessary libraries:

```python
import pandas as pd
from string import punctuation
import nltk
nltk.download('tagsets')
from nltk.data import load
nltk.download('averaged_perceptron_tagger')
from nltk import pos_tag
from nltk import word_tokenize
from collections import Counter
```

3. To see what different kinds of parts of speech `nltk` provides, add the following code:

```python
def get_tagsets():
    tagdict = load('help/tagsets/upenn_tagset.pickle')
    return list(tagdict.keys())

tag_list = get_tagsets()

print(tag_list)
```

The preceding code generates the following output:

```
['WDT', 'NNS', 'UH', 'PRP$', ':', 'EX', 'JJ', 'JJS', 'NNPS', 'TO', 'RBS', '.', 'JJR', 'WRB', 'SYM', 'RBR',
',', 'VBD', 'WP', ')', 'FW', 'PDT', 'VBG', 'VBZ', 'IN', "''", 'CC', '$', 'WP$', '``', 'VBN', 'POS', 'VB',
'RP', 'PRP', 'NN', 'CD', 'RB', 'LS', 'VBP', 'DT', 'NNP', '--', 'MD', '(']
```

Figure 2.16: List of PoS

4. Calculate the number of occurrences of each PoS by iterating through each document and annotating each word with the corresponding **pos** tag. Add the following code to implement this:

```
"""
This method will count the occurrence of pos
tags in each sentence.
"""
def get_pos_occurrence_freq(data, tag_list):
    # Get list of sentences in text_list
    text_list = data.text

    # create empty dataframe
    feature_df = pd.DataFrame(columns=tag_list)
    for text_line in text_list:

        # get pos tags of each word.
        pos_tags = [j for i, j in \
                    pos_tag(word_tokenize(text_line))]

        """
        create a dict of pos tags and their frequency
        in given sentence.
        """
        row = dict(Counter(pos_tags))
        feature_df = feature_df.append(row, ignore_index=True)
    feature_df.fillna(0, inplace=True)
    return feature_df
tag_list = get_tagsets()

data = pd.read_csv('../data/data.csv', header=0)
feature_df = get_pos_occurrence_freq(data, tag_list)
feature_df.head()
```

The preceding code generates the following output:

	WDT	NNS	UH	PRP$	:	EX	JJ	JJS	NNPS	TO	...	NN	CD	RB	LS	VBP	DT	NNP	--	MD	(
0	0.0	1.0	0.0	0.0	0.0	0.0	0.0	0.0	0.0	0.0	...	0.0	0.0	0.0	0.0	1.0	0.0	0.0	0.0	0.0	0.0
1	0.0	1.0	0.0	1.0	0.0	0.0	0.0	0.0	0.0	0.0	...	0.0	0.0	0.0	0.0	1.0	0.0	0.0	0.0	0.0	0.0
2	0.0	0.0	0.0	0.0	0.0	0.0	0.0	0.0	0.0	0.0	...	2.0	0.0	2.0	0.0	0.0	2.0	0.0	0.0	0.0	0.0
3	0.0	0.0	0.0	1.0	0.0	0.0	0.0	0.0	0.0	0.0	...	1.0	0.0	0.0	0.0	0.0	0.0	0.0	0.0	0.0	0.0
4	0.0	0.0	0.0	0.0	0.0	0.0	1.0	0.0	0.0	0.0	...	0.0	0.0	0.0	0.0	0.0	0.0	0.0	0.0	0.0	0.0

5 rows × 45 columns

Figure 2.17: Number of occurrences of each PoS in the sentence

5. To calculate the number of punctuation marks, add the following code:

```
def add_punctuation_count(feature_df, data):

    feature_df['num_of_unique_punctuations'] = data['text'].\
        apply(lambda x: len(set(x).intersection\
        (set(punctuation))))
    return feature_df


feature_df = add_punctuation_count(feature_df, data)

feature_df['num_of_unique_punctuations'].head()
```

The **add_punctuation_count()** method will find the intersection of the set of punctuation marks in the text and punctuation sets that were imported from the **string** module. Then, it will find the length of the intersection set in each row and add it to the **num_of_unique_punctuations** column of the DataFrame. The preceding code generates the following output:

```
0    0
1    0
2    1
3    1
4    0
Name:  num_of_unique_punctuations, dtype: int64
```

6. To calculate the number of capitalized words, add the following code:

```
def get_capitalized_word_count(feature_df, data):
    """
    The below code line will tokenize text in every row and
    create a set of only capital words, ten find the length of
    this set and add it to the column 'number_of_capital_words'
    of dataframe.
    """
    feature_df['number_of_capital_words'] = data['text'].\
        apply(lambda x: len([word for word in \
        word_tokenize(str(x)) if word[0].isupper()]))
    return feature_df

feature_df = get_capitalized_word_count(feature_df, data)

feature_df['number_of_capital_words'].head()
```

The preceding code will tokenize the text in every row and create a set of words consisting of only capital words. It will then find the length of this set and add it to the **number_of_capital_words** column of the DataFrame. The preceding code generates the following output:

```
0       1
1       1
2       1
3       1
4       1
Name:   number_of_capital_words, dtype: int64
```

The last line of the preceding code will print the **number_of_capital_words** column, which represents the count of the number of capital letter words in each row.

7. To calculate the number of lowercase words, add the following code:

```
def get_small_word_count(feature_df, data):
    """
    The below code line will tokenize text in every row and
    create a set of only small words, then find the length of
    this set and add it to the column 'number_of_small_words'
    of dataframe.
    """
    feature_df['number_of_small_words'] = data['text'].\
        apply(lambda x: len([word for word in \
        word_tokenize(str(x)) if word[0].islower()]))
    return feature_df

feature_df = get_small_word_count(feature_df, data)
feature_df['number_of_small_words'].head()
```

The preceding code will tokenize the text in every row and create a set of only small words, then find the length of this set and add it to the **number_of_small_words** column of the DataFrame. The preceding code generates the following output:

```
0      4
1      3
2      7
3      3
4      2
Name:  number_of_small_words, dtype: int64
```

The last line of the preceding code will print the **number_of_small_words** column, which represents the number of small letter words in each row.

8. To calculate the number of letters in the DataFrame, use the following code:

```
def get_number_of_alphabets(feature_df, data):

    feature_df['number_of_alphabets'] = data['text']. \
        apply(lambda x: len([ch for ch in str(x) \
        if ch.isalpha()]))
    return feature_df
```

```
feature_df = get_number_of_alphabets(feature_df, data)

feature_df['number_of_alphabets'].head()
```

The preceding code will break the text line into a list of characters in each row and add the count of that list to the **number_of_alphabets** columns. This will produce the following output:

```
0      19
1      18
2      28
3      14
4      13
Name:  number_of_alphabets, dtype: int64
```

The last line of the preceding code will print the **number_of_columns** column, which represents the count of the number of alphabets in each row.

9. To calculate the number of digits in the DataFrame, add the following code:

```
def get_number_of_digit_count(feature_df, data):
    """
    The below code line will break the text line in a list of
    digits in each row and add the count of that list into
    the columns 'number_of_digits'
    """

    feature_df['number_of_digits'] = data['text']. \
        apply(lambda x: len([ch for ch in str(x) \
        if ch.isdigit()]))
    return feature_df
feature_df = get_number_of_digit_count(feature_df, data)
feature_df['number_of_digits'].head()
```

The preceding code will get the digit count from each row and add the count of that list to the **number_of_digits** columns. The preceding code generates the following output:

```
0       0
1       0
2       0
3       0
4       0
Name:  number_of_digits, dtype: int64
```

10. To calculate the number of words in the DataFrame, add the following code:

```
def get_number_of_words(feature_df, data):
    """
    The below code line will break the text line in a list of
    words in each row and add the count of that list into
    the columns 'number_of_digits'
    """
    feature_df['number_of_words'] = data['text'].\
        apply(lambda x : len(word_tokenize(str(x))))

    return feature_df


feature_df = get_number_of_words(feature_df, data)
feature_df['number_of_words'].head()
```

The preceding code will split the text line into a list of words in each row and add the count of that list to the **number_of_digits** columns. We will get the following output:

```
0    5
1    4
2    9
3    5
4    3
Name:  number_of_words, dtype: int64
```

11. To calculate the number of whitespaces in the DataFrame, add the following code:

```
def get_number_of_whitespaces(feature_df, data):
    """
    The below code line will generate list of white spaces
    in each row and add the length of that list into
    the columns 'number_of_white_spaces
    """
    feature_df['number_of_white_spaces'] = data['text']. \
        apply(lambda x: len([ch for ch in str(x) \
        if ch.isspace()]))

    return feature_df
```

```
feature_df = get_number_of_whitespaces(feature_df, data)
feature_df['number_of_white_spaces'].head()
```

The preceding code will generate a list of whitespaces in each row and add the length of that list to the **number_of_white_spaces** columns. The preceding code generates the following output:

```
0       4
1       3
2       7
3       3
4       2
Name:  number_of_white_spaces, dtype: int64
```

12. To view the full feature set we have just created, add the following code:

```
feature_df.head()
```

We will be printing the head of the final DataFrame, which means we will print five rows of all the columns. We will get the following output:

	LS	TO	VBN	"	WP	UH	VBG	JJ	VBZ	--	...	CC	CD	POS	num_of_unique_punctuations	number_of_capit
0	0.0	0.0	0.0	0.0	1.0	0.0	0.0	0.0	0.0	0.0	...	0.0	0.0	0.0	0	1
1	0.0	0.0	0.0	0.0	0.0	0.0	0.0	0.0	0.0	0.0	...	0.0	0.0	0.0	0	1
2	0.0	0.0	0.0	0.0	0.0	0.0	0.0	0.0	0.0	0.0	...	0.0	0.0	0.0	1	1
3	0.0	0.0	0.0	0.0	1.0	0.0	0.0	0.0	1.0	0.0	...	0.0	0.0	0.0	1	1
4	0.0	0.0	0.0	0.0	0.0	0.0	0.0	1.0	1.0	0.0	...	0.0	0.0	0.0	0	1

5 rows × 52 columns

Figure 2.18: DataFrame consisting of the features we have created

With that, we have learned how to extract general features from the given text.

> **NOTE**
>
> To access the source code for this specific section, please refer to https://packt.live/3jSsLNh.
>
> You can also run this example online at https://packt.live/3hPFmPA.

Now, let's explore how we can extract unique features.

BAG OF WORDS (BOW)

The **Bag of Words (BoW)** model is one of the most popular methods for extracting features from raw texts.

In this technique, we convert each sentence into a vector. The length of this vector is equal to the number of unique words in all the documents. This is done in two steps:

1. The vocabulary or dictionary of all the words is generated.

2. The document is represented in terms of the presence or absence of all words.

A vocabulary or dictionary is created from all the unique possible words available in the corpus (all documents) and every single word is assigned a unique index number. In the second step, every document is represented by a list whose length is equal to the number of words in the vocabulary. The following exercise illustrates how BoW can be implemented using Python.

EXERCISE 2.13: CREATING A BAG OF WORDS

In this exercise, we will create a BoW representation for all the terms in a document and ascertain the 10 most frequent terms. In this exercise, we will use the `CountVectorizer` module from **sklearn**, which performs the following tasks:

- Tokenizes the collection of documents, also called a corpus

- Builds the vocabulary of unique words

- Converts a document into vectors using the previously built vocabulary

Follow these steps to implement this exercise:

1. Open a Jupyter Notebook.

2. Import the necessary libraries and declare a list corpus. Add the following code to implement this:

```
import pandas as pd

from sklearn.feature_extraction.text import CountVectorizer
```

3. Use the **CountVectorizer** function to create the BoW model. Add the following code to do this:

```
def vectorize_text(corpus):
    """
    Will return a dataframe in which every row will ,be
    vector representation of a document in corpus
    :param corpus: input text corpus
    :return: dataframe of vectors
    """
    bag_of_words_model = CountVectorizer()

    """
    performs the above described three tasks on
    the given data corpus.
    """
    dense_vec_matrix = bag_of_words_model.\
                        fit_transform(corpus).todense()
    bag_of_word_df = pd.DataFrame(dense_vec_matrix)
    bag_of_word_df.columns = sorted(bag_of_words_model.\
                            vocabulary_)
    return bag_of_word_df


corpus = ['Data Science is an overlap between Arts and Science',\
          'Generally, Arts graduates are right-brained and '\
          'Science graduates are left-brained',\
          'Excelling in both Arts and Science at a time '\
          'becomes difficult',\
          'Natural Language Processing is a part of Data Science']
df = vectorize_text(corpus)
df.head()
```

The **vectorize_text** method will take a document corpus as an argument and return a DataFrame in which every row will be a vector representation of a document in the corpus.

The preceding code generates the following output:

	an	and	are	arts	at	becomes	between	both	brained	data	...	language	left	natural	of	overlap	part	proces
0	1	1	0	1	0	0	1	0	0	1	...	0	0	0	0	1	0	0
1	0	1	2	1	0	0	0	0	2	0	...	0	1	0	0	0	0	0
2	0	1	0	1	1	1	0	1	0	0	...	0	0	0	0	0	0	0
3	0	0	0	0	0	0	0	0	0	1	...	1	0	1	1	0	1	1

4 rows × 26 columns

Figure 2.19: DataFrame of the output of the BoW model

4. Create a BoW model for the 10 most frequent terms. Add the following code to implement this:

```
def bow_top_n(corpus, n):
    """
    Will return a dataframe in which every row
    will be represented by presence or absence of top 10 most
    frequently occurring words in data corpus
    :param corpus: input text corpus
    :return: dataframe of vectors
    """
    bag_of_words_model_small = CountVectorizer(max_features=n)
    bag_of_word_df_small = pd.DataFrame\
    (bag_of_words_model_small.fit_transform\
    (corpus).todense())
    bag_of_word_df_small.columns = \
    sorted(bag_of_words_model_small.vocabulary_)
    return bag_of_word_df_small
df_2 = bow_top_n(corpus, 10)
df_2.head()
```

In the preceding code, we are checking the occurrence of the top 10 most frequent words in each sentence and creating a DataFrame out of it.

The preceding code generates the following output:

	an	and	are	arts	brained	data	graduates	is	right	science
0	1	1	0	1	0	1	0	1	0	2
1	0	1	2	1	2	0	2	0	1	1
2	0	1	0	1	0	0	0	0	0	1
3	0	0	0	0	0	1	0	1	0	1

Figure 2.20: DataFrame of the output of the BoW model for the 10 most frequent terms

NOTE

To access the source code for this specific section, please refer to https://packt.live/3gdhViJ.

You can also run this example online at https://packt.live/3hPUTi8.

In this section, we learned what BoW is and how to can use it to convert a sentence or document into a vector. BoW is the easiest way to convert text into a vector; however, it has a severe disadvantage. This method only considers the presence and absence of words in a sentence or document—not the frequency of the words/tokens in a document. If we are going to use the semantics of any sentence, the frequency of the words plays an important role. To overcome this issue, there is another feature extraction model called TFIDF, which we will discuss later in this chapter.

ZIPF'S LAW

According to Zipf's law, the number of times a word occurs in a corpus is inversely proportional to its rank in the frequency table. In simple terms, if the words in a corpus are arranged in descending order of their frequency of occurrence, then the frequency of the word at the ith rank will be proportional to $1/i$:

$$Frequency\ of\ a\ word\ =\ 1/rank\ of\ word\ in\ vocabulary$$

Figure 2.21: Zipf's law

This also means that the frequency of the most frequent word will be twice the frequency of the second most frequent word. For example, if we look at the Brown University Standard Corpus of Present-Day American English, the word "the" is the most frequent word (its frequency is 69,971), while the word "of" is the second most frequent (with a frequency of 36,411). As we can see, its frequency is almost half of the most frequently occurring word. To get a better understanding of this, let's perform a simple exercise.

EXERCISE 2.14: ZIPF'S LAW

In this exercise, we will plot both the expected and actual ranks and frequencies of tokens with the help of Zipf's law. We will be using the **20newsgroups** dataset provided by the **sklearn** library, which is a collection of newsgroup documents. Follow these steps to implement this exercise:

1. Open a Jupyter Notebook.

2. Import the necessary libraries:

```
from pylab import *
import nltk

nltk.download('stopwords')
from sklearn.datasets import fetch_20newsgroups
from nltk import word_tokenize
from nltk.corpus import stopwords
import matplotlib.pyplot as plt
import re
import string
from collections import Counter
```

Add two methods for loading stop words and the data from the **newsgroups_data_sample** variable:

```
def get_stop_words():
    stop_words = stopwords.words('english')
    stop_words = stop_words + list(string.printable)
    return stop_words
```

```
def get_and_prepare_data(stop_words):
    """
    This method will load 20newsgroups data and
    and remove stop words from it using given stop word list.
    :param stop_words:
    :return:
    """
    newsgroups_data_sample = \
    fetch_20newsgroups(subset='train')
    tokenized_corpus = [word.lower() for sentence in \
                        newsgroups_data_sample['data'] \
                        for word in word_tokenize\
                        (re.sub(r'([^\s\w]|_)+', ' ', sentence)) \
                        if word.lower() not in stop_words]
    return tokenized_corpus
```

In the preceding code, there are two methods; **get_stop_words()** will load stop word list from **nltk** data, while **get_and_prepare_data()** will load the **20newsgroups** data and remove stop words from it using the given stop word list.

3. Add the following method to calculate the frequency of each token:

```
def get_frequency(corpus, n):
    token_count_di = Counter(corpus)
    return token_count_di.most_common(n)
```

The preceding method uses the **Counter** class to count the frequency of tokens in the corpus and then return the most common **n** tokens.

4. Now, call all the preceding methods to calculate the frequency of the top 50 most frequent tokens:

```
stop_word_list = get_stop_words()
corpus = get_and_prepare_data(stop_word_list)
get_frequency(corpus, 50)
```

The preceding code generates the following output:

```
[('ax', 62412),
 ('edu', 21321),
 ('subject', 12265),
 ('com', 12134),
 ('lines', 11835),
 ('organization', 11233),
 ('one', 9017),
 ('would', 8910),
 ('writes', 7844),
 ('article', 7438),
 ('people', 5977),
 ('like', 5868),
 ('university', 5589),
 ('posting', 5507),
 ('know', 5134),
 ('get', 4998),
 ('host', 4996),
 ('nntp', 4814),
 ('max', 4776),
 ('think', 4583),
 ('also', 4308),
 ('use', 4187),
 ('time', 4102),
 ('new', 3986),
 ('good', 3759),
 ('ca', 3546),
 ('could', 3511),
 ('well', 3480),
 ('us', 3364),
 ('may', 3313),
 ('even', 3280),
 ('see', 3065),
 ('cs', 3041),
 ('two', 3015),
 ('way', 3002),
 ('god', 2998),
 ('first', 2976),
```

Figure 2.22: The 50 most frequent words of the corpus

5. Plot the actual ranks of words that we got from frequency dictionary and the ranks expected as per Zipf's law. Calculate the frequencies of the top 10,000 words using the preceding **get_frequency()** method and the expected frequencies of the same list using Zipf's law. For this, create two lists—an **actual_frequencies** and an **expected_frequencies** list. Use the log of actual frequencies to downscale the numbers. After getting the actual and expected frequencies, plot them using matplotlib:

```python
def get_actual_and_expected_frequencies(corpus):
    freq_dict = get_frequency(corpus, 1000)
    actual_frequencies = []
    expected_frequencies = []
    for rank, tup in enumerate(freq_dict):
        actual_frequencies.append(log(tup[1]))
        rank = 1 if rank == 0 else rank
        # expected frequency 1/rank as per zipf's law
        expected_frequencies.append(1 / rank)
    return actual_frequencies, expected_frequencies


def plot(actual_frequencies, expected_frequencies):
    plt.plot(actual_frequencies, 'g*', \
             expected_frequencies, 'ro')
    plt.show()


# We will plot the actual and expected frequencies
actual_frequencies, expected_frequencies = \
get_actual_and_expected_frequencies(corpus)
plot(actual_frequencies, expected_frequencies)
```

The preceding code generates the following output:

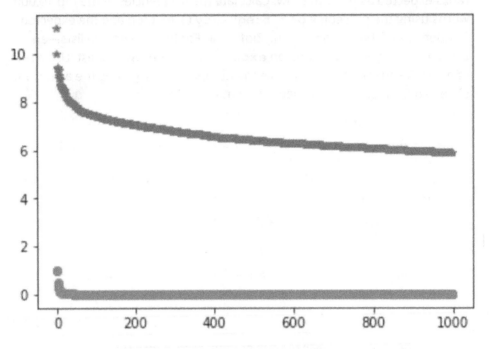

Figure 2.23: Illustration of Zipf's law

So, as we can see from the preceding output, both lines have almost the same slope. In other words, we can say that the lines (or graphs) depict the proportionality of two lists.

> **NOTE**
>
> To access the source code for this specific section, please refer to https://packt.live/30ZnKtD.
>
> You can also run this example online at https://packt.live/3f9ZFoT.

TERM FREQUENCY—INVERSE DOCUMENT FREQUENCY (TFIDF)

Term Frequency-Inverse Document Frequency (**TFIDF**) is another method of representing text data in a vector format. Here, once again, we'll represent each document as a list whose length is equal to the number of unique words/tokens in all documents (corpus), but the vector here not only represents the presence and absence of a word, but also the frequency of the word—both in the current document and the whole corpus.

This technique is based on the idea that the rarely occurring words are better representatives of the document than frequently occurring words. Hence, this representation gives more weightage to the rarer or less frequent words than frequently occurring words. It does so with the following formula:

$$Tf - idf = term - frequency * inverse\ document\ frequency$$

Figure 2.24: TFIDF formula

Here, term frequency is the frequency of a word in the given document. Inverse document frequency can be defined as log(D/df), where df is document frequency and D is the total number of documents in the background corpus.

Now, let's complete an exercise and learn how TFIDF can be implemented in Python.

EXERCISE 2.15: TFIDF REPRESENTATION

In this exercise, we will represent the input texts with their TFIDF vectors. We will use a **sklearn** module named **TfidfVectorizer**, which converts text into TFIDF vectors. Follow these steps to implement this exercise:

1. Open a Jupyter Notebook.

2. Import all the necessary libraries and create a method to calculate the TFIDF of the corpus. Add the following code to implement this:

```
from sklearn.feature_extraction.text import TfidfVectorizer

def get_tf_idf_vectors(corpus):
    tfidf_model = TfidfVectorizer()
    vector_list = tfidf_model.fit_transform(corpus).todense()
    return vector_list
```

3. To create a TFIDF model, write the following code:

```
corpus = ['Data Science is an overlap between Arts and Science',\
          'Generally, Arts graduates are right-brained and '\
          'Science graduates are left-brained',\
          'Excelling in both Arts and Science at a '\
          'time becomes difficult',\
          'Natural Language Processing is a part of Data Science']
vector_list = get_tf_idf_vectors(corpus)
print(vector_list)
```

In the preceding code, the **get_tf_idf_vectors()** method will generate TFIDF vectors from the corpus. You will then call this method on a given corpus. The preceding code generates the following output:

```
[[0.40332811 0.25743911 0.         0.25743911 0.         0.
  0.40332811 0.         0.         0.31798852 0.         0.
  0.         0.         0.         0.31798852 0.         0.
  0.         0.         0.40332811 0.         0.         0.
  0.42094668 0.                    ]
 [0.         0.159139   0.49864399 0.159139   0.         0.
  0.         0.         0.49864399 0.         0.         0.
  0.24932199 0.49864399 0.         0.         0.         0.24932199
  0.         0.         0.         0.         0.         0.24932199
  0.13010656 0.                    ]
 [0.         0.22444946 0.         0.22444946 0.35164346 0.35164346
  0.         0.35164346 0.         0.         0.35164346 0.35164346
  0.         0.         0.35164346 0.         0.         0.
  0.         0.         0.         0.         0.         0.
  0.18350214 0.35164346]
 [0.         0.         0.         0.         0.         0.
  0.         0.         0.         0.30887228 0.         0.
  0.         0.         0.         0.30887228 0.39176533 0.
  0.39176533 0.39176533 0.         0.39176533 0.39176533 0.
  0.2044394  0.                  ]]
```

Figure 2.25: TFIDF representation of the 10 most frequent terms

The preceding output represents the TFIDF vectors for each row. As you can see from the results, each document is represented by a list whose length is equal to the unique words in the corpus and in each list (vector). The vector contains the TFIDF values of the words at their corresponding index.

> **NOTE**
>
> To access the source code for this specific section, please refer to https://packt.live/3gdzsHA.
>
> You can also run this example online at https://packt.live/3fdP5gS.

In the next section, we will solve an activity to extract specific features from texts.

FINDING TEXT SIMILARITY – APPLICATION OF FEATURE EXTRACTION

So far in this chapter, we have learned how to generate vectors from text. These vectors are then fed to machine learning algorithms to perform various tasks. Other than using them in machine learning applications, we can also perform simple NLP tasks using these vectors. Finding the string similarity is one of them. This is a technique in which we find the similarity between two strings by converting them into vectors. The technique is mainly used in full-text searching.

There are different techniques for finding the similarity between two strings or texts. They are explained one by one here:

- **Cosine similarity**: The cosine similarity is a technique to find the similarity between the two vectors by calculating the cosine of the angle between them. As we know, the cosine of a zero-degree angle is 1 (meaning the cosine similarity of two identical vectors is 1), while the cosine of 180 degrees is -1 (meaning the cosine of two opposite vectors is -1). Thus, we can use this cosine angle to find the similarity between the vectors from 1 to -1. To use this technique in finding text similarity, we convert text into vectors using one of the previously discussed techniques and find the similarity between the vectors of the text. This is calculated as follows:

$$Similarity = cos(\theta) = cos(A,B) = A.B \ / \ (|A| * |B|)$$

Figure 2.26: Cosine similarity

Here, A and B are two vectors, A.B is the dot product of two vectors, and |A| and |B| are the magnitude of two vectors.

- **Jaccard similarity**: This is another technique that's used to calculate the similarity between the two texts, but it only works on BoW vectors. The Jaccard similarity is calculated as the ratio of the number of terms that are common between two text documents to the total number of unique terms present in those texts.

Consider the following example. Suppose there are two texts:

Text 1: I like detective Byomkesh Bakshi.

Text 2: Byomkesh Bakshi is not a detective; he is a truth seeker.

The common terms are "Byomkesh," "Bakshi," and "detective."

The number of common terms in the texts is three.

The unique terms present in the texts are "I," "like," "is," "not," "a," "he," "is," "truth," and "seeker." So, the number of unique terms is nine.

Therefore, the Jaccard similarity is 3/9 = 0.3.

To get a better understanding of text similarity, we will complete an exercise.

EXERCISE 2.16: CALCULATING TEXT SIMILARITY USING JACCARD AND COSINE SIMILARITY

In this exercise, we will calculate the Jaccard and cosine similarity for a given pair of texts. Follow these steps to complete this exercise:

1. Open a Jupyter Notebook.

2. Insert a new cell and add the following code to import the necessary packages:

```
from nltk import word_tokenize
from nltk.stem import WordNetLemmatizer
from sklearn.feature_extraction.text import TfidfVectorizer
from sklearn.metrics.pairwise import cosine_similarity
lemmatizer = WordNetLemmatizer()
```

3. Create a function to extract the Jaccard similarity between a pair of sentences by adding the following code:

```
def extract_text_similarity_jaccard(text1, text2):
    """
    This method will return Jaccard similarity between two texts
    after lemmatizing them.
    :param text1: text1
    :param text2: text2
    :return: similarity measure
    """
    lemmatizer = WordNetLemmatizer()

    words_text1 = [lemmatizer.lemmatize(word.lower()) \
                   for word in word_tokenize(text1)]
    words_text2 = [lemmatizer.lemmatize(word.lower()) \
                   for word in word_tokenize(text2)]
    nr = len(set(words_text1).intersection(set(words_text2)))
    dr = len(set(words_text1).union(set(words_text2)))
    jaccard_sim = nr / dr
    return jaccard_sim
```

4. Declare three variables named **pair1**, **pair2**, and **pair3**, as follows.

```
pair1 = ["What you do defines you", "Your deeds define you"]
pair2 = ["Once upon a time there lived a king.", \
         "Who is your queen?"]
pair3 = ["He is desperate", "Is he not desperate?"]
```

5. To check the Jaccard similarity between the statements in **pair1**, write the following code:

```
extract_text_similarity_jaccard(pair1[0],pair1[1])
```

The preceding code generates the following output:

```
0.14285714285714285
```

6. To check the Jaccard similarity between the statements in **pair2**, write the following code:

```
extract_text_similarity_jaccard(pair2[0],pair2[1])
```

The preceding code generates the following output:

```
0.0
```

7. To check the Jaccard similarity between the statements in **pair3**, write the following code:

```
extract_text_similarity_jaccard(pair3[0],pair3[1])
```

The preceding code generates the following output:

```
0.6
```

8. To check the cosine similarity, use the **TfidfVectorizer()** method to get the vectors of each text:

```
def get_tf_idf_vectors(corpus):
    tfidf_vectorizer = TfidfVectorizer()
    tfidf_results = tfidf_vectorizer.fit_transform(corpus).\
                    todense()
    return tfidf_results
```

9. Create a corpus as a list of texts and get the TFIDF vectors of each text document. Add the following code to do this:

```
corpus = [pair1[0], pair1[1], pair2[0], \
          pair2[1], pair3[0], pair3[1]]
tf_idf_vectors = get_tf_idf_vectors(corpus)
```

10. To check the cosine similarity between the initial two texts, write the following code:

```
cosine_similarity(tf_idf_vectors[0],tf_idf_vectors[1])
```

The preceding code generates the following output:

```
array([[0.3082764]])
```

11. To check the cosine similarity between the third and fourth texts, write the following code:

```
cosine_similarity(tf_idf_vectors[2],tf_idf_vectors[3])
```

The preceding code generates the following output:

```
array([[0.]])
```

12. To check the cosine similarity between the fifth and sixth texts, write the following code:

```
cosine_similarity(tf_idf_vectors[4],tf_idf_vectors[5])
```

The preceding code generates the following output:

```
array([[0.80368547]])
```

So, in this exercise, we learned how to check the similarity between texts. As you can see, the texts **"He is desperate"** and **"Is he not desperate?"** returned similarity results of 0.80 (meaning they are highly similar), whereas sentences such as **"Once upon a time there lived a king."** and **"Who is your queen?"** returned zero as their similarity measure.

> **NOTE**
>
> To access the source code for this specific section, please refer to https://packt.live/2Eyw0JC.
>
> You can also run this example online at https://packt.live/2XbGRQ3.

WORD SENSE DISAMBIGUATION USING THE LESK ALGORITHM

The Lesk algorithm is used for resolving word sense disambiguation. Suppose we have a sentence such as "On the bank of river Ganga, there lies the scent of spirituality" and another sentence, "I'm going to withdraw some cash from the bank". Here, the same word—that is, "bank"—is used in two different contexts. For text processing results to be accurate, the context of the words needs to be considered.

In the Lesk algorithm, words with ambiguous meanings are stored in the background in **synsets**. The definition that is closer to the meaning of a word being used in the context of the sentence will be taken as the right definition. Let's perform a simple exercise to get a better idea of how we can implement this.

EXERCISE 2.17: IMPLEMENTING THE LESK ALGORITHM USING STRING SIMILARITY AND TEXT VECTORIZATION

In this exercise, we are going to implement the Lesk algorithm step by step using the techniques we have learned so far. We will find the meaning of the word "bank" in the sentence, "On the banks of river Ganga, there lies the scent of spirituality." We will use cosine similarity as well as Jaccard similarity here. Follow these steps to complete this exercise:

1. Open a Jupyter Notebook.

2. Insert a new cell and add the following code to import the necessary libraries:

```
import pandas as pd
from sklearn.metrics.pairwise import cosine_similarity
from nltk import word_tokenize
from sklearn.feature_extraction.text import TfidfVectorizer
from sklearn.datasets import fetch_20newsgroups
import numpy as np
```

3. Define a method for getting the TFIDF vectors of a corpus:

```
def get_tf_idf_vectors(corpus):
    tfidf_vectorizer = TfidfVectorizer()
    tfidf_results = tfidf_vectorizer.fit_transform\
                    (corpus).todense()
    return tfidf_results
```

4. Define a method to convert the corpus into lowercase:

```
def to_lower_case(corpus):
    lowercase_corpus = [x.lower() for x in corpus]
    return lowercase_corpus
```

5. Define a method to find the similarity between the sentence and the possible definitions and return the definition with the highest similarity score:

```
def find_sentence_definition(sent_vector,defnition_vectors):
    """
    This method will find cosine similarity of sentence with
    the possible definitions and return the one with
    highest similarity score along with the similarity score.
    """
    result_dict = {}
```

```
        for definition_id,def_vector in definition_vectors.items():
            sim = cosine_similarity(sent_vector,def_vector)
            result_dict[definition_id] = sim[0][0]
        definition  = sorted(result_dict.items(), \
                               key=lambda x: x[1], \
                               reverse=True)[0]
        return definition[0],definition[1]
```

6. Define a corpus with random sentences with the sentence and the two definitions as the top three sentences:

```
corpus = ["On the banks of river Ganga, there lies the scent "\
          "of spirituality",\
          "An institute where people can store extra "\
          "cash or money.",\
          "The land alongside or sloping down to a river or lake"
          "What you do defines you",\
          "Your deeds define you",\
          "Once upon a time there lived a king.",\
          "Who is your queen?",\
          "He is desperate",\
          "Is he not desperate?"]
```

7. Use the previously defined methods to find the definition of the word bank:

```
lower_case_corpus  = to_lower_case(corpus)
corpus_tf_idf  = get_tf_idf_vectors(lower_case_corpus)
sent_vector = corpus_tf_idf[0]
definition_vectors = {'def1':corpus_tf_idf[1],\
                       'def2':corpus_tf_idf[2]}
definition_id, score = \
find_sentence_definition(sent_vector,definition_vectors)
print("The definition of word {} is {} with similarity of {}".\
      format('bank',definition_id,score))
```

You will get the following output:

```
The definition of word bank is def2 with similarity of
0.14419130686278897
```

As we already know, **def2** represents a riverbank. So, we have found the correct definition of the word here. In this exercise, we have learned how to use text vectorization and text similarity to find the right definition of ambiguous words.

> **NOTE**
>
> To access the source code for this specific section, please refer to https://packt.live/39GzJAs.
>
> You can also run this example online at https://packt.live/3fbxQwK.

WORD CLOUDS

Unlike numeric data, there are very few ways in which text data can be represented visually. The most popular way of visualizing text data is by using word clouds. A word cloud is a visualization of a text corpus in which the sizes of the tokens (words) represent the number of times they have occurred, as shown in the following image:

Figure 2.27: Example of a word cloud

In the following exercise, we will be using a Python library called **wordcloud** to build a word cloud from the **20newsgroups** dataset.

Let's go through an exercise to understand this better.

EXERCISE 2.18: GENERATING WORD CLOUDS

In this exercise, we will visualize the most frequently occurring words in the first 1,000 articles from **sklearn**'s **fetch_20newsgroups** text dataset using a word cloud. Follow these steps to complete this exercise:

1. Open a Jupyter Notebook.

2. Import the necessary libraries and dataset. Add the following code to do this:

```
import nltk
nltk.download('stopwords')
import matplotlib.pyplot as plt
plt.rcParams['figure.dpi'] = 200
from sklearn.datasets import fetch_20newsgroups
from nltk.corpus import stopwords
from wordcloud import WordCloud
import matplotlib as mpl
mpl.rcParams['figure.dpi'] = 200
```

3. Write the **get_data()** method to fetch the data:

```
def get_data(n):
    newsgroups_data_sample = fetch_20newsgroups(subset='train')
    text = str(newsgroups_data_sample['data'][:n])
    return text
```

4. Add a method to remove stop words:

```
def load_stop_words():
    other_stopwords_to_remove = ['\\n', 'n', '\\', '>', \
                                  'nLines', 'nI',"n'"]
    stop_words = stopwords.words('english')
    stop_words.extend(other_stopwords_to_remove)
    stop_words = set(stop_words)
    return stop_words
```

5. Add the **generate_word_cloud()** method to generate a word cloud object:

```
def generate_word_cloud(text, stopwords):
    """
    This method generates word cloud object
    with given corpus, stop words and dimensions
    """

    wordcloud = WordCloud(width = 800, height = 800, \
                          background_color ='white', \
                          max_words=200, \
                          stopwords = stopwords, \
                          min_font_size = 10).generate(text)
    return wordcloud
```

6. Get 1,000 documents from the **20newsgroup** data, get the stop word list, generate a word cloud object, and finally plot the word cloud with matplotlib:

```
text = get_data(1000)
stop_words = load_stop_words()
wordcloud = generate_word_cloud(text, stop_words)
plt.imshow(wordcloud, interpolation='bilinear')
plt.axis("off")
plt.show()
```

The preceding code generates the following output:

Figure 2.28: Word cloud representation of the first 10 articles

So, in this exercise, we learned what word clouds are and how to generate word clouds with Python's **wordcloud** library and visualize this with matplotlib.

> **NOTE**
>
> To access the source code for this specific section, please refer to https://packt.live/30eaSRn.
>
> You can also run this example online at https://packt.live/2EzqLJJ.

In the next section, we will explore other visualizations, such as dependency parse trees and named entities.

OTHER VISUALIZATIONS

Apart from word clouds, there are various other ways of visualizing texts. Some of the most popular ways are listed here:

- **Visualizing sentences using a dependency parse tree**: Generally, the phrases constituting a sentence depend on each other. We depict these dependencies by using a tree structure known as a dependency parse tree. For instance, the word "*helps*" in the sentence "God helps those who help themselves" depends on two other words. These words are "*God*" (the one who helps) and "*those*" (the ones who are helped).

- **Visualizing named entities in a text corpus**: In this case, we extract the named entities from texts and highlight them by using different colors.

Let's go through the following exercise to understand this better.

EXERCISE 2.19: OTHER VISUALIZATIONS DEPENDENCY PARSE TREES AND NAMED ENTITIES

In this exercise, we will look at two of the most popular visualization methods, after word clouds, which are dependency parse trees and using named entities. Follow these steps to complete this exercise:

1. Open a Jupyter Notebook.

2. Insert a new cell and add the following code to import the necessary libraries:

```
import spacy
from spacy import displacy
!python -m spacy download
en_core_web_sm
import en_core_web_sm
nlp = en_core_web_sm.load()
```

3. Depict the sentence "God helps those who help themselves" using a dependency parse tree with the following code:

```
doc = nlp('God helps those who help themselves')
displacy.render(doc, style='dep', jupyter=True)
```

The preceding code generates the following output:

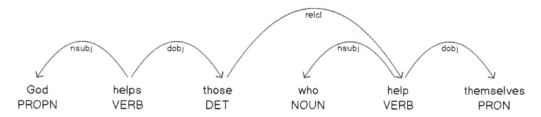

Figure 2.29: Dependency parse tree

4. Visualize the named entities of the text corpus by adding the following code:

```
text = 'Once upon a time there lived a saint named '\
        'Ramakrishna Paramahansa. His chief disciple '\
        'Narendranath Dutta also known as Swami Vivekananda '\
        'is the founder of Ramakrishna Mission and '\
        'Ramakrishna Math.'
doc2 = nlp(text)
displacy.render(doc2, style='ent', jupyter=True)
```

The preceding code generates the following output:

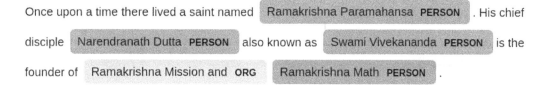

Figure 2.30: Named entities

> **NOTE**
>
> To access the source code for this specific section, please refer to https://packt.live/313m4iD.
>
> You can also run this example online at https://packt.live/3103fgr.

Now that you have learned about visualizations, we will solve an activity based on them to gain an even better understanding.

ACTIVITY 2.02: TEXT VISUALIZATION

In this activity, you will create a word cloud for the 50 most frequent words in a dataset. The dataset we will use consists of random sentences that are not clean. First, we need to clean them and create a unique set of frequently occurring words.

> **NOTE**
>
> The text_corpus.txt file that's being used in this activity can be found at https://packt.live/2DiVIBj.

Follow these steps to implement this activity:

1. Import the necessary libraries.

2. Fetch the dataset.

3. Perform the preprocessing steps, such as text cleaning, tokenization, and lemmatization, on the fetched data.

4. Create a set of unique words along with their frequencies for the 50 most frequently occurring words.

5. Create a word cloud for these top 50 words.

6. Justify the word cloud by comparing it with the word frequency that you calculated.

> **NOTE**
>
> The solution to this activity can be found on page 375.

SUMMARY

In this chapter, you have learned about various types of data and ways to deal with unstructured text data. Text data is usually extremely noisy and needs to be cleaned and preprocessed, which mainly consists of tokenization, stemming, lemmatization, and stop-word removal. After preprocessing, features are extracted from texts using various methods, such as BoW and TFIDF. These methods convert unstructured text data into structured numeric data. New features are created from existing features using a technique called feature engineering. In the last part of this chapter, we explored various ways of visualizing text data, such as word clouds.

In the next chapter, you will learn how to develop machine learning models to classify texts using the feature extraction methods you have learned about in this chapter. Moreover, different sampling techniques and model evaluation parameters will be introduced.

3

DEVELOPING A TEXT CLASSIFIER

OVERVIEW

This chapter starts with an introduction to the various types of machine learning methods, that is, the supervised and unsupervised methods. You will learn about hierarchical clustering and k-means clustering and implement them using various datasets. Next, you will explore tree-based methods such as random forest and XGBoost. Finally, you will implement an end-to-end text classifier in order to categorize text on the basis of its content.

INTRODUCTION

In the previous chapters, you learned about various extraction methods, such as tokenization, stemming, lemmatization, and stop-word removal, which are used to extract features from unstructured text. We also discussed Bag of Words and **Term Frequency-Inverse Document Frequency** (**TFIDF**).

In this chapter, you will learn how to use these extracted features to develop machine learning models. These models are capable of solving real-world problems, such as detecting whether sentiments carried by texts are positive or negative, predicting whether emails are spam or not, and so on. We will also cover concepts such as supervised and unsupervised learning, classifications and regressions, sampling and splitting data, along with evaluating the performance of a model in depth. This chapter also discusses how to load and save these models for future use.

MACHINE LEARNING

Machine learning is the scientific study of algorithms and statistical models that computer systems use to perform a specific task without using explicit instructions, relying on patterns and inference instead.

Machine learning algorithms are fed with large amounts of data that they can work on to build a model. This model is later used by businesses to generate solutions that help them analyze data and build strategies for the future. For example, a beverage production company can make use of multiple datasets to better understand the trends of their product's consumption over the course of a year. This would help them reduce wastage and better predict the requirements of their consumers. Machine learning is further categorized into **unsupervised** and **supervised** learning. Let's explore these two terms in detail.

UNSUPERVISED LEARNING

Unsupervised learning is the method by which algorithms learn patterns within data that is not labeled. Since labels (supervisors) are absent, it is referred to as unsupervised learning. In unsupervised learning, you provide the algorithm with the feature data and it learns patterns from the data on its own.

Unsupervised learning is further classified into clustering and association:

- **Clustering**: Clustering is the process of combining objects into groups called clusters. For example, if there are 50 students who need to be categorized based on their attributes, we do not use any specific attribute(s) to create segments. Rather, we try to learn the hidden patterns that exist in their attributes and categorize them accordingly. This process is known as cluster analysis or clustering (one of the most popular types of unsupervised learning). When handed a set of text documents, we can divide them into groups that are similar with the help of clustering. A common example of clustering could be when you search for a term on Google and similar pages or links are recommended. These recommendations are powered by document clustering.

- **Association**: Another type of unsupervised learning is association rule mining. We use association rule mining to obtain groups of items that occur together frequently. The most common use case of association rule mining is to identify customers' buying patterns. For example, in a supermarket, customers who tend to buy milk and bread generally tend to buy cheese. This information can be used to design supermarket layouts. An application of association rule mining in **Natural Language Processing** (**NLP**) is to find similar words; for example, *outstanding, excellent*, and *superb* are all synonyms of *good*. Association rule mining can easily find patterns like this in any NLP dataset. However, the detailed theoretical explanations of these algorithms are beyond the scope of this chapter.

Let's explore the different types of clustering. In particular, we will be talking about hierarchical and k-means clustering, and the different scenarios in which they should be used. However, before we dive into those, it's important to understand the concept of distance metrics, which is what we use to create clusters and identify similar data points. The most common distance metric is Euclidean, which is calculated as follows:

$$Euclidean\ Distance(p,\ q) = \sqrt{\sum_{i=1}^{n} (p_i - q_i)^2}$$

Figure 3.1: Formula for Euclidean distance

In the case of machine learning, *p* and *q* are different data points in the dataset and p_i, q_i are the different features of those data points.

HIERARCHICAL CLUSTERING

Hierarchical clustering algorithms group similar objects together to create a cluster with the help of a **dendrogram**. In this algorithm, we can vary the number of clusters as per our requirements. First, we construct a matrix consisting of distances between each pair of instances (data points). After that, we construct a **dendrogram** (a representation of clusters in the form of a tree) based on the distances between them. We truncate the tree at a location corresponding to the number of clusters we need.

For example, imagine that you have 10 documents and want to group them into a number of categories based on their attributes (the number of words they contain, the number of paragraphs, punctuation, and so on) and don't have any fixed number of categories in mind. This is a use case of hierarchical clustering. Let's assume that we have a dataset containing the features of the 10 documents. Firstly, the distances between each pair of documents from the set of 10 documents are calculated. After that, we construct a **dendrogram** and truncate it at a suitable position to get a suitable number of clusters:

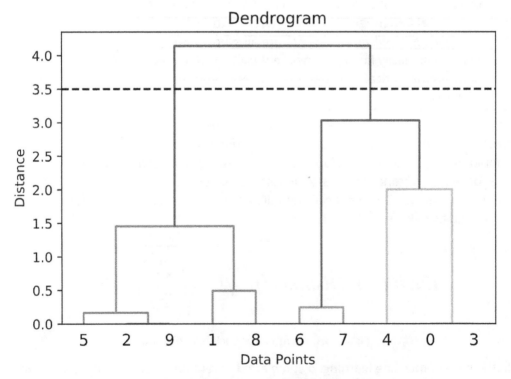

Figure 3.2: Output dendrogram after performing hierarchical clustering

In the preceding graph, we can perform a truncation at distance 3.5 to get two clusters or at 2.5 to get three clusters, depending on the requirements. To create a dendrogram using scikit-learn, we can use the following code:

```
import scipy.cluster.hierarchy as sch

dendrogram = sch.dendrogram(sch.linkage(X, method='ward'))

plt.title('Dendrogram')
plt.show()
```

Here, **X** is the dataset that we want to perform hierarchical clustering with. Let's perform an exercise to understand how we can implement this.

EXERCISE 3.01: PERFORMING HIERARCHICAL CLUSTERING

In this exercise, we will analyze the text documents in sklearn's **fetch_20newsgroups** dataset. The 20 newsgroups dataset contains news articles on 20 different topics. We will make use of hierarchical clustering to classify the documents into different groups. Once the clusters have been created, we will compare them with their actual categories. Follow these steps to implement this exercise:

1. Open a Jupyter Notebook.

2. Insert a new cell and add the following code to import the necessary libraries:

```
from sklearn.datasets import fetch_20newsgroups
from scipy.cluster.hierarchy import ward, dendrogram
import matplotlib as mpl
from scipy.cluster.hierarchy import fcluster
from sklearn.metrics.pairwise import cosine_similarity
import pandas as pd
import numpy as np
import matplotlib.pyplot as plt
%matplotlib inline
import re
import string
from nltk import word_tokenize
from nltk.corpus import stopwords
from nltk.stem import WordNetLemmatizer
from sklearn.feature_extraction.text import TfidfVectorizer
from collections import Counter
```

```
from pylab import *
import nltk
import warnings
warnings.filterwarnings('ignore')
```

3. Download a list of stop words and the **Wordnet** corpus from **nltk**. Insert a new cell and add the following code to implement this:

```
nltk.download('stopwords')
stop_words=stopwords.words('english')
stop_words=stop_words+list(string.printable)
nltk.download('wordnet')
lemmatizer=WordNetLemmatizer()
```

4. Specify the categories of news articles we want to fetch to perform our clustering task. We will use three categories: "For sale", "Electronics", and "Religion". Add the following code to do this:

```
categories= ['misc.forsale', 'sci.electronics', \
             'talk.religion.misc']
```

5. To fetch the dataset, add the following lines of code:

```
news_data = fetch_20newsgroups(subset='train', \
                               categories=categories, \
                               shuffle=True, random_state=42, \
                               download_if_missing=True)
```

6. To view the data of the fetched content, add the following code:

```
news_data['data'][:5]
```

The preceding code generates the following output:

```
['From: Steve@Busop.cit.wayne.edu (Steve Teolis)\nSubject: Re: *** TurboGrafx System For SALE ***\nOrganiz
ation: Wayne State University\nLines: 38\nDistribution: na\nNNTP-Posting-Host: 141.217.75.24\n\n>TurboGraf
x-16 Base Unit (works like new) with:\n>          1 Controller\n>          AC Adapter\n>          Antenna hookup\n>
* Games:\n>          Kieth Courage\n>          Victory Run\n>          Fantasy Zone\n>          Military Madne
ss\n>          Battle Royal\n>          Legendary Axe\n>          Blazing Lasers\n>          Bloody Wolf\n>\n>
-------------------------------------\n>* Will sell games separatley at $25 each\n>     --------------------
-------------------\n\nYour kidding, $210.00, man o man, you can buy the system new for $49.00 at \nElectro
nic Boutique and those games are only about $15 - $20.00 brand new.  \nMaybe you should think about that p
rice again if you REALLY need the money.\n\n\n\n\n\n\n                    \n                                \n
-=-=-=-=-=-=-=-=-=-=-=-=-\n                              Wayne State University    \n
\n                    Steve Teolis                \n                                    6050 Cass Ave. #
238        \n                    Detroit, MI  48202        \n
\n                    Steve@Busop.cit.wayne.edu     \n                         -=-=-=-=-=-=-=-=-=-=-=
-=-=-=-\n',
```

Figure 3.3: The first five news articles

7. To check the categories of news articles, insert a new cell and add the following code:

```
print(news_data.target)
```

The target is the variable that we predict by making use of the rest of the variables in a dataset. The preceding code generates the following output:

```
[0 0 1 … 0 1 0]
```

Here, **0** refers to **misc.forsale**, **1** refers to **sci.electronics**, and **2** refers to **talk.religion.misc**.

8. To store **news_data** and the corresponding categories in a pandas **DataFrame** and view it, write the following code:

```
news_data_df = pd.DataFrame({'text' : news_data['data'], \
                             'category': news_data.target})
news_data_df.head()
```

The preceding code generates the following output:

	text	category
0	From: Steve@Busop.cit.wayne.edu (Steve Teolis)...	0
1	From: jks2x@holmes.acc.Virginia.EDU (Jason K. ...	0
2	From: wayne@uva386.schools.virginia.edu (Tony ...	1
3	From: lihan@ccwf.cc.utexas.edu (Bruce G. Bostw...	1
4	From: myoakam@cis.ohio-state.edu (micah r yoak...	0

Figure 3.4: Text corpus of news data corresponding to the categories in a DataFrame

9. To count the number of occurrences of each category appearing in this dataset, write the following code:

```
news_data_df['category'].value_counts()
```

The preceding code generates the following output:

```
1        591
0        585
2        377
Name:  category, dtype: int64
```

10. Use a lambda function to extract tokens from each "text" of the **news_data_df** DataFrame. Check whether any of these tokens is a stop word, lemmatize the ones that are not stop words, and then concatenate them to recreate the sentence. Make use of the **join** function to concatenate a list of words into a single sentence. To replace anything other than letters, digits, and whitespaces with blank space, use a regular expression (**re**). Add the following code to do this:

```
news_data_df['cleaned_text'] = news_data_df['text']\
                      .apply(lambda x : ' '.join\
                      ([lemmatizer.lemmatize\
                       (word.lower())\
                      for word in word_tokenize\
                      (re.sub(r'([^\s\w]|_)+', ' ',\
                       str(x))) if word.lower() \
                      not in stop_words]))
```

11. Create a TFIDF matrix and transform it into a DataFrame. Add the following code to do this:

```
tfidf_model = TfidfVectorizer(max_features=200)
tfidf_df = pd.DataFrame(tfidf_model.fit_transform\
            (news_data_df['cleaned_text']).todense())
tfidf_df.columns = sorted(tfidf_model.vocabulary_)
tfidf_df.head()
```

The preceding code generates the following output:

	00	10	100	12	14	15	16	20	25	30	...	well	wire	wiring	without	word	work	world	would
0	0.435655	0.0	0.000000	0.0	0.000000	0.127775	0.136811	0.127551	0.133311	0.0	...	0.0	0.0	0.0	0.0	0.0	0.113042	0.000000	0.000000
1	0.000000	0.0	0.000000	0.0	0.000000	0.294937	0.000000	0.000000	0.000000	0.0	...	0.0	0.0	0.0	0.0	0.0	0.000000	0.000000	0.000000
2	0.000000	0.0	0.000000	0.0	0.000000	0.000000	0.000000	0.000000	0.000000	0.0	...	0.0	0.0	0.0	0.0	0.0	0.000000	0.000000	0.000000
3	0.000000	0.0	0.000000	0.0	0.000000	0.000000	0.000000	0.000000	0.000000	0.0	...	0.0	0.0	0.0	0.0	0.0	0.000000	0.142267	0.106317
4	0.000000	0.0	0.207003	0.0	0.191897	0.182138	0.000000	0.000000	0.000000	0.0	...	0.0	0.0	0.0	0.0	0.0	0.000000	0.000000	0.000000

5 rows × 200 columns

Figure 3.5: TFIDF representation as a DataFrame

12. Calculate the distance using the sklearn library:

```
from sklearn.metrics.pairwise import \
euclidean_distances as euclidean
dist = 1 - euclidean(tfidf_df)
```

13. Now, create a dendrogram for the TFIDF representation of documents:

```
import scipy.cluster.hierarchy as sch
dendrogram = sch.dendrogram(sch.linkage(dist, method='ward'))
plt.xlabel('Data Points')
plt.ylabel('Euclidean Distance')
plt.title('Dendrogram')
plt.show()
```

The preceding code generates the following output:

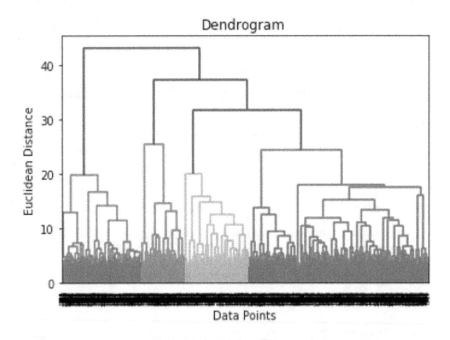

Figure 3.6: Truncated dendrogram

Here, you can see that a cluster count of four seems optimal.

14. Use the **fcluster()** function to obtain the cluster labels of the clusters that were obtained by hierarchical clustering:

```
k=4
clusters = fcluster(sch.linkage(dist, method='ward'), k, \
                    criterion='maxclust')
clusters
```

The preceding code generates the following output:

```
array([3, 3, 3, ..., 4, 4, 1], dtype=int32)
```

15. Make use of the **crosstab** function of pandas to compare the clusters we have obtained with the actual categories of news articles. Add the following code to implement this:

```
news_data_df['obtained_clusters'] = clusters
pd.crosstab(news_data_df['category']\
            .replace({0:'misc.forsale', \
                      1:'sci.electronics', \
                      2:'talk.religion.misc'}),\
           news_data_df['obtained_clusters']\
            .replace({1 : 'cluster_1', 2 : 'cluster_2', \
                      3 : 'cluster_3', 4: 'cluster_4'}))
```

The preceding code generates the following output:

obtained_clusters	cluster_1	cluster_2	cluster_3	cluster_4
category				
misc.forsale	133	1	98	353
sci.electronics	140	0	442	9
talk.religion.misc	74	232	71	0

Figure 3.7: Crosstab between actual categories and obtained clusters

Using the preceding image, we can analyze the high-level patterns that the clustering algorithm found to group the articles into one of the four clusters. As you can see, cluster 2 has mostly religion-related articles, while cluster 3 consists of primarily sales-related articles. The other two clusters do not have a proper distinction. The reason for this could be that the model figured out that words related to "religion" and "for sale" appeared frequently in the articles that were classified into those respective clusters, while the articles on "electronics" consist of mostly generic words.

> **NOTE**
>
> To access the source code for this specific section, please refer to https://packt.live/39A4wyL.
>
> You can also run this example online at https://packt.live/3ge4ezQ.

One major disadvantage of hierarchical clustering is scalability. Using hierarchical clustering for large datasets is very difficult; for such cases, we can use k-means clustering. Let us explore how this works.

K-MEANS CLUSTERING

In this algorithm, we segregate the given instances (data points) into "k" number of groups (here, k is a natural number). First, we choose k centroids. We assign each instance to its nearest centroid, thereby creating k groups. This is the assignment phase, which is followed by the update phase.

In the update phase, new centroids for each of these k groups are calculated. The data points are reassigned to their nearest newly calculated centroids. The assignment phase and the update phase are carried on repeatedly until the assignment of data points no longer changes.

For example, suppose you have 10 documents. You want to group them into three categories based on their attributes, such as the number of words they contain, the number of paragraphs, punctuation, and the tone of the document. In this case, we will assume that k is 3; that is, we want to create these three groups. Firstly, three centroids need to be chosen. In the initialization phase, each of these 10 documents is assigned to one of these three categories, thereby forming three groups. In the update phase, the centroids of the three newly formed groups are calculated. To decide the optimal number of clusters (that is, k), we execute k-means clustering for various values of k and note down their performances (sum of squared errors). We try to select a small value for k that has the lowest sum of squared errors. This method is called the **elbow method**.

The scikit-learn library can be used to perform k-means in Python using the following code:

```
from sklearn.cluster import KMeans
kmeans = KMeans(n_clusters=4)
kmeans.fit(X)
clusters = kmeans.predict(X)
```

Here, we create the base model using the **kmeans** class of scikit-learn. Then, we train the model using the **fit** function. The trained model can then be used to get clusters using the predict function, where **X** represents a DataFrame of independent variables. Let's perform an exercise to get a better understanding of k-means clustering.

EXERCISE 3.02: IMPLEMENTING K-MEANS CLUSTERING

In this exercise, we will create four clusters from text documents in sklearn's **fetch_20newsgroups** text dataset using k-means clustering. We will compare these clusters with the actual categories and use the elbow method to obtain the optimal number of clusters. Follow these steps to implement this exercise:

1. Open a Jupyter Notebook.

2. Insert a new cell and add the following code to import the necessary packages:

```
import pandas as pd
from sklearn.datasets import fetch_20newsgroups
import matplotlib.pyplot as plt
%matplotlib inline
import re
import string
from nltk import word_tokenize
from nltk.corpus import stopwords
from nltk.stem import WordNetLemmatizer
from sklearn.feature_extraction.text import TfidfVectorizer
from collections import Counter
from pylab import *
import nltk
nltk.download('stopwords')
nltk.download('punkt')
nltk.download('wordnet')
import warnings
warnings.filterwarnings('ignore')
import seaborn as sns
sns.set()
import numpy as np
from scipy.spatial.distance import cdist
from sklearn.cluster import KMeans
```

3. To use stop words for the English language and the **WordNet** corpus for lemmatization, add the following code:

```
stop_words = stopwords.words('english')
stop_words = stop_words + list(string.printable)
lemmatizer = WordNetLemmatizer()
```

4. To specify the categories of news articles, add the following code:

```
categories= ['misc.forsale', 'sci.electronics', \
             'talk.religion.misc']
```

5. Use the following lines of code to fetch the dataset and store it in a pandas DataFrame:

```
news_data = fetch_20newsgroups(subset='train', \
                               categories=categories, \
                               shuffle=True, \
                               random_state=42, \
                               download_if_missing=True)
news_data_df = pd.DataFrame({'text' : news_data['data'], \
                             'category': news_data.target})
```

6. Use the lambda function to extract tokens from each "**text**" of the **news_data_df** DataFrame. Discard the tokens if they're stop words, lemmatize them if they're not, and then concatenate them to recreate the sentence. Use the join function to concatenate a list of words into a single sentence and use the regular expression method (**re**) to replace anything other than alphabets, digits, and whitespaces with a blank space. Add the following code to do this:

```
news_data_df['cleaned_text'] = news_data_df['text']\
                 .apply(lambda x : ' '.join\
                 ([lemmatizer.lemmatize(word.lower()) \
                 for word in word_tokenize\
                 (re.sub(r'([^\s\w]|_)+', ' ', \
                         str(x))) \
                 if word.lower() not in stop_words]))
```

7. Use the following lines of code to create a TFIDF matrix and transform it into a DataFrame:

```
tfidf_model = TfidfVectorizer(max_features=200)
tfidf_df = pd.DataFrame(tfidf_model\
                .fit_transform\
                (news_data_df['cleaned_text']).todense())
tfidf_df.columns = sorted(tfidf_model.vocabulary_)
tfidf_df.head()
```

The preceding code generates the following output:

	00	10	100	12	14	15	16	20	25	30	...	well	wire	wiring	without	word
0	0.435655	0.0	0.000000	0.0	0.000000	0.127775	0.136811	0.127551	0.133311	0.0	...	0.0	0.0	0.0	0.0	0.0
1	0.000000	0.0	0.000000	0.0	0.000000	0.294937	0.000000	0.000000	0.000000	0.0	...	0.0	0.0	0.0	0.0	0.0
2	0.000000	0.0	0.000000	0.0	0.000000	0.000000	0.000000	0.000000	0.000000	0.0	...	0.0	0.0	0.0	0.0	0.0
3	0.000000	0.0	0.000000	0.0	0.000000	0.000000	0.000000	0.000000	0.000000	0.0	...	0.0	0.0	0.0	0.0	0.0
4	0.000000	0.0	0.207003	0.0	0.191897	0.182138	0.000000	0.000000	0.000000	0.0	...	0.0	0.0	0.0	0.0	0.0

5 rows × 200 columns

Figure 3.8: TFIDF representation as a DataFrame

8. Use the **KMeans** function of sklearn to create four clusters from a TFIDF representation of news articles. Add the following code to do this:

```
kmeans = KMeans(n_clusters=4)
kmeans.fit(tfidf_df)
y_kmeans = kmeans.predict(tfidf_df)
news_data_df['obtained_clusters'] = y_kmeans
```

9. Use pandas' **crosstab** function to compare the clusters we have obtained with the actual categories of the news articles. Add the following code to do this:

```
pd.crosstab(news_data_df['category']\
            .replace({0:'misc.forsale', \
                      1:'sci.electronics', \
                      2:'talk.religion.misc'}),\
            news_data_df['obtained_clusters']\
            .replace({0 : 'cluster_1',\
                      1 : 'cluster_2', 2 : 'cluster_3', \
                      3: 'cluster_4'}))
```

The preceding code generates the following output:

obtained_clusters	cluster_1	cluster_2	cluster_3	cluster_4
category				
misc.forsale	133	1	98	353
sci.electronics	140	0	442	9
talk.religion.misc	74	232	71	0

Figure 3.9: Crosstab between the actual categories and obtained clusters xxx

From the figure above, you can see, cluster 2 has majorly religion related articles and cluster 4 has mostly for sale related articles. The other two clusters do now have a proper distinction but cluster 3 has majority of the electronic articles.

10. Finally, to obtain the optimal value of k (that is, the number of clusters), execute the k-means algorithm for values of k ranging from **1** to **6**. For each value of **k**, store the distortion—that is, the mean of the distances of the documents from their nearest cluster center. Look for the value of k where the slope of the plot changes rapidly. Add the following code to implement this:

```
distortions = []
K = range(1,6)
for k in K:
    kmeanModel = KMeans(n_clusters=k)
    kmeanModel.fit(tfidf_df)
    distortions.append(sum(np.min(cdist\
    (tfidf_df, kmeanModel.cluster_centers_, \
     'euclidean'), axis=1)) / tfidf_df.shape[0])

plt.plot(K, distortions, 'bx-')
plt.xlabel('k')
plt.ylabel('Distortion')
plt.title('The Elbow Method showing the optimal number '\
          'of clusters')
plt.show()
```

The preceding code generates the following output:

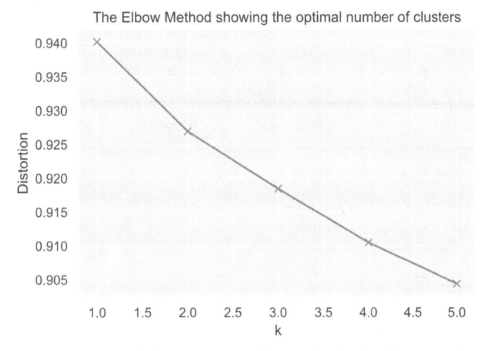

Figure 3.10: Optimal clusters represented in a graph using the elbow method

From the preceding graph, we can conclude that the optimal number of clusters is 2.

> **NOTE**
>
> To access the source code for this specific section, please refer to https://packt.live/2EuZckB.
>
> You can also run this example online at https://packt.live/333x6Hw.

We have seen how unsupervised learning can be implemented in Python. Now, let us talk about supervised learning.

SUPERVISED LEARNING

Unlike unsupervised learning, supervised learning algorithms need labeled data. They learn how to automatically generate labels or predict values by analyzing various features of the data provided. For example, say you have already starred important text messages on your phone, and you want to automate the task of going through all your messages daily (considering they are important and marked already). This is a use case for supervised learning. Here, messages that have been starred previously can be used as labeled data. Using this data, you can create two types of models that are capable of the following:

- Classifying whether new messages are important

- Predicting the probability of new messages being important

The first type is called classification, while the second type is called regression. Let's learn about classification first.

CLASSIFICATION

Say you have two types of food, of which type 1 tastes sweet and type 2 tastes salty, and you need to determine how an unknown food will taste using various attributes of the food (such as color, fragrance, shape, and ingredients). This is an instance of classification.

Here, the two classes are class 1, which tastes sweet, and class 2, which tastes salty. The features that are used in this classification are color, fragrance, the ingredients used to prepare the dish, and so on. These features are called independent variables. The class (sweet or salty) is called a dependent variable.

Formally, classification algorithms are those that learn patterns from a given dataset to determine classes of unknown datasets. Some of the most widely used classification algorithms are logistic regression, Naive Bayes, k-nearest neighbor, and tree methods. Let's learn about each of them.

LOGISTIC REGRESSION

Despite having the term "regression" in it, logistic regression is used for probabilistic classification. In this case, the dependent variable (the outcome) is binary, which means that the values can be represented by 0 or 1. For example, consider that you need to decide whether an email is spam or not. Here, the value of the decision (the dependent variable, or the outcome) can be considered to be 1 if the email is spam; otherwise, it will be 0. No other outcome is possible. The independent variables (that is, the features) will consist of various attributes of the email, such as the number of occurrences of certain keywords and so on. We can then make use of the logistic regression algorithm to create a model that predicts if the email is spam (1) or not (0), as shown in the following graph:

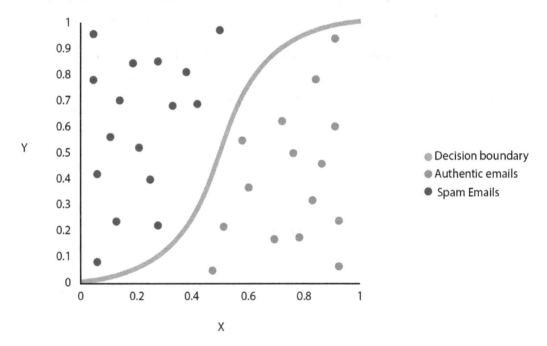

Figure 3.11: Example of logistic regression

Here, the decision boundary is created by training a logistic regression model that helps us classify spam emails.

The scikit-learn library can be used to perform logistic regression in Python using the following code:

```
from sklearn.linear_model import LogisticRegression
log_reg = LogisticRegression()
log_reg.fit(X,y)
predicted_labels = log_reg.predict(X)
predicted_probability = log_reg.predict_proba(X)[:,1]
```

Here, we create the base model using the **LogisticRegression** class of scikit-learn. Then, we train the model using the **fit** function. The trained model can then be used to make predictions, and we can also get probability estimates for each class using the **predict_proba** function. Here, **X** represents a DataFrame of independent variables, whereas **y** represents a DataFrame of dependent variables.

EXERCISE 3.03: TEXT CLASSIFICATION — LOGISTIC REGRESSION

In this exercise, we will classify reviews of musical instruments on Amazon with the help of the logistic regression classification algorithm.

> **NOTE**
>
> To download the dataset, visit https://packt.live/3hQ6UEe.

Follow these steps to implement this exercise:

1. Open a Jupyter Notebook.

2. Insert a new cell and add the following code to import the necessary packages:

```
import pandas as pd
import matplotlib.pyplot as plt
%matplotlib inline
import re
import string
from nltk import word_tokenize
from nltk.stem import WordNetLemmatizer
from sklearn.feature_extraction.text import TfidfVectorizer
from collections import Counter
from pylab import *
import nltk
nltk.download('punkt')
```

```
nltk.download('wordnet')
import warnings
warnings.filterwarnings('ignore')
```

3. Read the data file in JSON format using pandas. Add the following code to implement this:

```
review_data = pd.read_json\
                ('data/reviews_Musical_Instruments_5.json', \
                lines=True)
review_data[['reviewText', 'overall']].head()
```

The preceding code generates the following output:

	reviewText	overall
0	Not much to write about here, but it does exac...	5
1	The product does exactly as it should and is q...	5
2	The primary job of this device is to block the...	5
3	Nice windscreen protects my MXL mic and preven...	5
4	This pop filter is great. It looks and perform...	5

Figure 3.12: Data stored in a DataFrame

4. Use a **lambda** function to extract tokens from each **'reviewText'** of this DataFrame, lemmatize them, and concatenate them side by side. Use the **join** function to concatenate a list of words into a single sentence. Use the regular expression method (**re**) to replace anything other than alphabetical characters, digits, and whitespaces with blank space. Add the following code to implement this:

```
lemmatizer = WordNetLemmatizer()
review_data['cleaned_review_text'] = review_data['reviewText']\
                                    .apply(lambda x : ' '.join\
                                    ([lemmatizer.lemmatize\
                                    (word.lower()) \
                                    for word in word_tokenize\
                                    (re.sub(r'([^\s\w]|_)+', ' ',\
                                    str(x)))]))
```

5. Create a DataFrame from the TFIDF matrix representation of the cleaned version of **reviewText**. Add the following code to implement this:

```
review_data[['cleaned_review_text', 'reviewText', \
             'overall']].head()
```

The preceding code generates the following output:

	cleaned_review_text	reviewText	overall
0	not much to write about here but it doe exactl...	Not much to write about here, but it does exac...	5
1	the product doe exactly a it should and is qui...	The product does exactly as it should and is q...	5
2	the primary job of this device is to block the...	The primary job of this device is to block the...	5
3	nice windscreen protects my mxl mic and preven...	Nice windscreen protects my MXL mic and preven...	5
4	this pop filter is great it look and performs ...	This pop filter is great. It looks and perform...	5

Figure 3.13: Review texts before and after cleaning, along with their overall scores

6. Create a TFIDF matrix and transform it into a **DataFrame**. Add the following code to do this:

```
tfidf_model = TfidfVectorizer(max_features=500)
tfidf_df = pd.DataFrame(tfidf_model.fit_transform\
           (review_data['cleaned_review_text']).todense())
tfidf_df.columns = sorted(tfidf_model.vocabulary_)
tfidf_df.head()
```

The preceding code generates the following output:

	10	100	12	20	34	able	about	accurate	acoustic	actually	...	won	work	worked	worth	would	wrong	year	yet	you
0	0.0	0.0	0.0	0.0	0.0	0.000000	0.159684	0.0	0.0	0.0	...	0.0	0.134327	0.0	0.0	0.000000	0.0	0.0	0.0	0.000000
1	0.0	0.0	0.0	0.0	0.0	0.000000	0.000000	0.0	0.0	0.0	...	0.0	0.085436	0.0	0.0	0.000000	0.0	0.0	0.0	0.067074
2	0.0	0.0	0.0	0.0	0.0	0.000000	0.000000	0.0	0.0	0.0	...	0.0	0.000000	0.0	0.0	0.115312	0.0	0.0	0.0	0.079880
3	0.0	0.0	0.0	0.0	0.0	0.339573	0.000000	0.0	0.0	0.0	...	0.0	0.000000	0.0	0.0	0.000000	0.0	0.0	0.0	0.000000
4	0.0	0.0	0.0	0.0	0.0	0.000000	0.000000	0.0	0.0	0.0	...	0.0	0.000000	0.0	0.0	0.000000	0.0	0.0	0.0	0.303608

5 rows × 500 columns

Figure 3.14: A TFIDF representation as a DataFrame

7. The following lines of code are used to create a new column target, which will have **0** if the **overall** parameter is less than **4**, and 1 otherwise. Add the following code to implement this:

```
review_data['target'] = review_data['overall'].apply\
                    (lambda x : 0 if x<=4 else 1)
review_data['target'].value_counts()
```

The preceding code generates the following output:

```
1       6938
0       3323
Name: target, dtype: int64
```

8. Use sklearn's **LogisticRegression()** function to fit a logistic regression model on the TFIDF representation of these reviews after cleaning them. Add the following code to implement this:

```
from sklearn.linear_model import LogisticRegression
logreg = LogisticRegression()
logreg.fit(tfidf_df,review_data['target'])
predicted_labels = logreg.predict(tfidf_df)
logreg.predict_proba(tfidf_df)[:,1]
```

The preceding code generates the following output:

```
array([0.57146961, 0.68579907, 0.56068939, ..., 0.65979968, \
       0.5495679 , 0.21186011])
```

9. Use the **crosstab** function of pandas to compare the results of our classification model with the actual classes (**'target'**, in this case) of the reviews. Add the following code to do this:

```
review_data['predicted_labels'] = predicted_labels
pd.crosstab(review_data['target'], \
            review_data['predicted_labels'])
```

The preceding code generates the following output:

predicted_labels	0	1
target		
0	1543	1780
1	626	6312

Figure 3.15: Crosstab between actual target values and predicted labels

Here, we can see **1543** instances with the target label 0 that are correctly classified and **1780** such instances that are wrongly classified. Furthermore, **6312** instances with the target label 1 are correctly classified, whereas **626** such instances are wrongly classified.

> **NOTE**
>
> To access the source code for this specific section, please refer to https://packt.live/3hOaKxJ.
>
> You can also run this example online at https://packt.live/309yKWc.

We've seen how to implement logistic regression; now, let's look at Naïve Bayes classification.

NAIVE BAYES CLASSIFIERS

Just like logistic regression, a Naive Bayes classifier is another kind of probabilistic classifier. It is based on Bayes' theorem, which is shown here:

$$P(A/B) = \frac{P(B/A)P(A)}{P(B)}$$

Figure 3.16: Bayes' theorem

In the preceding formula, A and B are events and P(B) is not equal to 0. P(A/B) is the probability of event A occurring, given that event B is true. Similarly, P(B/A) is the probability of event B occurring, given that event A is true. P(B) is the probability of the occurrence of event B.

Say there is an online platform where hotel customers can provide a review for the service provided to them. The hotel now wants to figure out whether new reviews on the platform are appreciative in nature or not. Here, P(A) = the probability of the review being an appreciative one, while P(B) = the probability of the review being long. Now, we've come across a review that is long and want to figure out the probability of it being appreciative. To do that, we need to calculate P(A/B). P(B/A) will be the probability of appreciative reviews being long. From the training dataset, we can easily calculate P(B/A), P(A), and P(B) and then use Bayes' theorem to calculate P(A/B).

Similar to logistic regression, the scikit-learn library can be used to perform naïve Bayes classification and can be implemented in Python using the following code:

```
from sklearn.naive_bayes import GaussianNB
nb = GaussianNB()
nb.fit(X,y)
predicted_labels = nb.predict(X)
predicted_probability = nb.predict_proba(X)[:,1]
```

Here, we created the base model using the **GaussianNB** class of scikit-learn. Then, we trained the model using the **fit** function. The trained model can then be used to make predictions; we can also get probability estimates for each class using the **predict_proba** function. Here, **X** represents a DataFrame of independent variables, whereas **y** represents a DataFrame of dependent variables.

EXERCISE 3.04: TEXT CLASSIFICATION – NAIVE BAYES

In this exercise, we will classify reviews of musical instruments on Amazon with the help of the Naïve Bayes classification algorithm. Follow these steps to implement this exercise:

> **NOTE**
>
> To download the dataset for this exercise, visit https://packt.live/3hQ6UEe.

1. Open a Jupyter Notebook.

2. Insert a new cell and add the following code to import the necessary packages:

```
import pandas as pd
import matplotlib.pyplot as plt
%matplotlib inline
import re
import string
from nltk import word_tokenize
from nltk.stem import WordNetLemmatizer
from sklearn.feature_extraction.text import TfidfVectorizer
from collections import Counter
from pylab import *
import nltk
nltk.download('punkt')
nltk.download('wordnet')
import warnings
warnings.filterwarnings('ignore')
```

3. Read the data file in JSON format using pandas. Add the following code to implement this:

```
review_data = pd.read_json\
                ('data/reviews_Musical_Instruments_5.json', \
                lines=True)
review_data[['reviewText', 'overall']].head()
```

The preceding code generates the following output:

	reviewText	overall
0	Not much to write about here, but it does exac...	5
1	The product does exactly as it should and is q...	5
2	The primary job of this device is to block the...	5
3	Nice windscreen protects my MXL mic and preven...	5
4	This pop filter is great. It looks and perform...	5

Figure 3.17: Data stored in a DataFrame

4. Use a **lambda** function to extract tokens from each **'reviewText'** of this DataFrame, lemmatize them, and concatenate them side by side. Use the **join** function to concatenate a list of words into a single sentence. Use the regular expression method (**re**) to replace anything other than alphabets, digits, and whitespaces with blank space. Add the following code to implement this:

```
lemmatizer = WordNetLemmatizer()
review_data['cleaned_review_text'] = review_data['reviewText']\
                                .apply(lambda x : ' '.join\
                                ([lemmatizer.lemmatize\
                                (word.lower()) \
                                for word in word_tokenize\
                                (re.sub(r'([^\s\w]|_)+', ' ',\
                                str(x)))]))
```

5. Create a DataFrame from the TFIDF matrix representation of the cleaned version of **reviewText**. Add the following code to implement this:

```
review_data[['cleaned_review_text', 'reviewText', \
            'overall']].head()
```

The preceding code generates the following output:

	cleaned_review_text	reviewText	overall
0	not much to write about here but it doe exactl...	Not much to write about here, but it does exac...	5
1	the product doe exactly a it should and is qui...	The product does exactly as it should and is q...	5
2	the primary job of this device is to block the...	The primary job of this device is to block the...	5
3	nice windscreen protects my mxl mic and preven...	Nice windscreen protects my MXL mic and preven...	5
4	this pop filter is great it look and performs ...	This pop filter is great. It looks and perform...	5

Figure 3.18: Review texts before and after cleaning, along with their overall scores

6. Create a TFIDF matrix and transform it into a DataFrame. Add the following code to do this:

```
tfidf_model = TfidfVectorizer(max_features=500)
tfidf_df = pd.DataFrame(tfidf_model.fit_transform\
            (review_data['cleaned_review_text']).todense())
tfidf_df.columns = sorted(tfidf_model.vocabulary_)
tfidf_df.head()
```

The preceding code generates the following output:

	10	100	12	20	34	able	about	accurate	acoustic	actually	...	won	work	worked	worth	would	wrong	year	yet	you
0	0.0	0.0	0.0	0.0	0.0	0.000000	0.159684	0.0	0.0	0.0	...	0.0	0.134327	0.0	0.0	0.000000	0.0	0.0	0.0	0.000000
1	0.0	0.0	0.0	0.0	0.0	0.000000	0.000000	0.0	0.0	0.0	...	0.0	0.085436	0.0	0.0	0.000000	0.0	0.0	0.0	0.067074
2	0.0	0.0	0.0	0.0	0.0	0.000000	0.000000	0.0	0.0	0.0	...	0.0	0.000000	0.0	0.0	0.115312	0.0	0.0	0.0	0.079880
3	0.0	0.0	0.0	0.0	0.0	0.339573	0.000000	0.0	0.0	0.0	...	0.0	0.000000	0.0	0.0	0.000000	0.0	0.0	0.0	0.000000
4	0.0	0.0	0.0	0.0	0.0	0.000000	0.000000	0.0	0.0	0.0	...	0.0	0.000000	0.0	0.0	0.000000	0.0	0.0	0.0	0.303608

5 rows × 500 columns

Figure 3.19: A TFIDF representation as a DataFrame

7. The following lines of code are used to create a new column target, which will have the value **0** if the **'overall'** parameter is less than **4**, and 1 otherwise. Add the following code to implement this:

```
review_data['target'] = review_data['overall']\
                        .apply(lambda x : 0 if x<=4 else 1)
review_data['target'].value_counts()
```

The preceding code generates the following output:

```
1    6938
0    3323
Name: target, dtype: int64
```

8. Use sklearn's **GaussianNB()** function to fit a Gaussian Naive Bayes model on the TFIDF representation of these reviews after cleaning them. Add the following code to do this:

```
from sklearn.naive_bayes import GaussianNB
nb = GaussianNB()
nb.fit(tfidf_df,review_data['target'])
predicted_labels = nb.predict(tfidf_df)
nb.predict_proba(tfidf_df)[:,1]
```

The preceding code generates the following output:

```
array([[9.97730158e-01, 3.63599675e-09, 9.45692105e-07, ...,
        2.46001047e-02, 3.43660991e-08, 1.72767906e-27])
```

The preceding screenshot shows the predicted probabilities of the input **tfidf_df** dataset.

9. Use the **crosstab** function of pandas to compare the results of our classification model with the actual classes (**'target'**, in this case) of the reviews. Add the following code to do this:

```
review_data['predicted_labels'] = predicted_labels
pd.crosstab(review_data['target'], \
            review_data['predicted_labels'])
```

The preceding code generates the following output:

predicted_labels	0	1
target		
0	2333	990
1	2380	4558

Figure 3.20: Crosstab between actual target values and predicted labels

Here, we can see **2333** instances with the target label 0 that are correctly classified and **990** such instances that have been wrongly classified. Furthermore, **4558** instances with the target label 1 have been correctly classified, whereas **2380** such instances have been wrongly classified.

> **NOTE**
>
> To access the source code for this specific section, please refer to https://packt.live/2DnoeBx.
>
> You can also run this example online at https://packt.live/3fcjT1t.

We'll explore k-nearest neighbors in the next section.

K-NEAREST NEIGHBORS

k-nearest neighbors is an algorithm that can be used to solve both regression and classification. In this chapter, we will focus on the classification aspect of the algorithm as it is used for NLP applications. Consider, for instance, the saying "birds of a feather flock together." This means that people who have similar interests prefer to stay close to each other and form groups. This characteristic is called **homophily**. This characteristic is the main idea behind the k-nearest neighbors classification algorithm.

To classify an unknown object, k number of other objects located nearest to it with class labels will be looked into. The class that occurs the most among them will be assigned to it—that is, the object with an unknown class. The value of k is chosen by running experiments on the training dataset to find the most optimal value. When dealing with text data for a given document, we interpret "nearest neighbors" as other documents that are the most similar to the unknown document.

We can make use of the scikit-learn library to implement the k-nearest neighbors algorithm in Python using the following code:

```
from sklearn.neighbors import KNeighborsClassifier
knn = KNeighborsClassifier(n_neighbors=3)
knn.fit(X,y)
prediction = knn.predict(X)
```

Here, we created the base model using the **KNeighborsClassifier** class of scikit-learn and pass it the value of k, which in this case is 3. Then, we trained the model using the **fit** function. The trained model can then be used to make predictions using the **predict** function. **X** represents a DataFrame of independent variables, whereas **y** represents a DataFrame of dependent variables.

Now that we have an understanding of different types of classification, let's see how we can implement them.

EXERCISE 3.05: TEXT CLASSIFICATION USING THE K-NEAREST NEIGHBORS METHOD

In this exercise, we will classify reviews of musical instruments on Amazon with the help of the k-nearest neighbors classification algorithm. Follow these steps to implement this exercise:

> **NOTE**
>
> To download the dataset for this exercise, visit https://packt.live/3hQ6UEe.

1. Open a Jupyter Notebook.

2. Insert a new cell and add the following code to import the necessary packages:

```
import pandas as pd
import matplotlib.pyplot as plt
%matplotlib inline
import re
import string
from nltk import word_tokenize
from nltk.stem import WordNetLemmatizer
from sklearn.feature_extraction.text import TfidfVectorizer
from collections import Counter
from pylab import *
import nltk
nltk.download('punkt')
nltk.download('wordnet')
import warnings
warnings.filterwarnings('ignore')
```

3. Read the data file in JSON format using pandas. Add the following code to implement this:

```
review_data = pd.read_json\
                ('data/reviews_Musical_Instruments_5.json',\
                 lines=True)
review_data[['reviewText', 'overall']].head()
```

The preceding code generates the following output:

	reviewText	overall
0	Not much to write about here, but it does exac...	5
1	The product does exactly as it should and is q...	5
2	The primary job of this device is to block the...	5
3	Nice windscreen protects my MXL mic and preven...	5
4	This pop filter is great. It looks and perform...	5

Figure 3.21: Data stored in a DataFrame

4. Use a **lambda** function to extract tokens from each **reviewText** of this DataFrame, lemmatize them, and concatenate them side by side. Use the **join** function to concatenate a list of words into a single sentence. Use the regular expression method (**re**) to replace anything other than alphabets, digits, and whitespaces with blank space. Add the following code to implement this:

```
lemmatizer = WordNetLemmatizer()
review_data['cleaned_review_text'] = review_data['reviewText']\
                                    .apply(lambda x : ' '.join\
                                    ([lemmatizer.lemmatize\
                                      (word.lower()) \
                                    for word in word_tokenize\
                                    (re.sub(r'([^\s\w]|_)+', ' ',\
                                    str(x)))]))
```

5. Create a DataFrame from the TFIDF matrix representation of the cleaned version of **reviewText**. Add the following code to implement this:

```
review_data[['cleaned_review_text', 'reviewText', \
            'overall']].head()
```

The preceding code generates the following output:

	cleaned_review_text	reviewText	overall
0	not much to write about here but it doe exactl...	Not much to write about here, but it does exac...	5
1	the product doe exactly a it should and is qui...	The product does exactly as it should and is q...	5
2	the primary job of this device is to block the...	The primary job of this device is to block the...	5
3	nice windscreen protects my mxl mic and preven...	Nice windscreen protects my MXL mic and preven...	5
4	this pop filter is great it look and performs ...	This pop filter is great. It looks and perform...	5

Figure 3.22: Review texts before and after cleaning, along with their overall scores

6. Create a TFIDF matrix and transform it into a DataFrame. Add the following code to do this:

```
tfidf_model = TfidfVectorizer(max_features=500)
tfidf_df = pd.DataFrame(tfidf_model.fit_transform\
            (review_data['cleaned_review_text']).todense())
tfidf_df.columns = sorted(tfidf_model.vocabulary_)
tfidf_df.head()
```

The preceding code generates the following output:

	10	100	12	20	34	able	about	accurate	acoustic	actually	...	won	work	worked	worth	would	wrong	year	yet	you
0	0.0	0.0	0.0	0.0	0.0	0.000000	0.159684	0.0	0.0	0.0	...	0.0	0.134327	0.0	0.0	0.000000	0.0	0.0	0.0	0.000000
1	0.0	0.0	0.0	0.0	0.0	0.000000	0.000000	0.0	0.0	0.0	...	0.0	0.085436	0.0	0.0	0.000000	0.0	0.0	0.0	0.067074
2	0.0	0.0	0.0	0.0	0.0	0.000000	0.000000	0.0	0.0	0.0	...	0.0	0.000000	0.0	0.0	0.115312	0.0	0.0	0.0	0.079880
3	0.0	0.0	0.0	0.0	0.0	0.339573	0.000000	0.0	0.0	0.0	...	0.0	0.000000	0.0	0.0	0.000000	0.0	0.0	0.0	0.000000
4	0.0	0.0	0.0	0.0	0.0	0.000000	0.000000	0.0	0.0	0.0	...	0.0	0.000000	0.0	0.0	0.000000	0.0	0.0	0.0	0.303608

5 rows × 500 columns

Figure 3.23: A TFIDF representation as a DataFrame

7. The following lines of code are used to create a new column target, which will have **0** if the **overall** parameter is less than **4**, and 1 otherwise. Add the following code to implement this:

```
review_data['target'] = review_data['overall']\
                        .apply(lambda x : 0 if x<=4 else 1)
review_data['target'].value_counts()
```

The preceding code generates the following output:

```
1       6938
0       3323
Name:   target, dtype: int64
```

8. Use sklearn's **KNeighborsClassifier()** function to fit a three-nearest neighbor model on the TFIDF representation of these reviews after cleaning them. Use the **crosstab** function of pandas to compare the results of our classification model with the actual classes (**'target'**, in this case) of the reviews:

```
from sklearn.neighbors import KNeighborsClassifier
knn = KNeighborsClassifier(n_neighbors=3)
knn.fit(tfidf_df,review_data['target'])
review_data['predicted_labels_knn'] = knn.predict(tfidf_df)
pd.crosstab(review_data['target'], \
            review_data['predicted_labels_knn'])
```

The preceding code generates the following output:

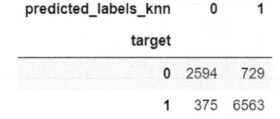

predicted_labels_knn	0	1
target		
0	2594	729
1	375	6563

Figure 3.24: Crosstab between actual target values and predicted
labels by k-nearest neighbors

Here, we can see **2594** instances with the target label as 0 correctly classified and **729** such instances wrongly classified. Furthermore, **6563** instances with the target label as 1 are correctly classified, whereas **375** such instances are wrongly classified. You have just learned how to perform text classification with the help of various classification algorithms.

> **NOTE**
>
> To access the source code for this specific section, please refer to https://packt.live/338XQqb.
>
> You can also run this example online at https://packt.live/39E5zNW.

In the next section, you will learn about regression, which is another type of supervised learning.

REGRESSION

To better understand regression, consider a practical example. For example, say you have photos of several people, along with a list of their respective ages, and you need to predict the ages of some other people from their photos. This is a use case for regression.

In the case of regression, the dependent variable (age, in this example) is continuous. The independent variables—that is, features—consist of the attributes of the images, such as the color intensity of each pixel. Formally, regression analysis refers to the process of learning a mapping function, which relates features or predictors (inputs) to the dependent variable (output).

There are various types of regression: **univariate, multivariate, simple, multiple, linear, non-linear, polynomial regression, stepwise regression, ridge regression, lasso regression**, and **elastic net regression**. If there is just one dependent variable, then it is referred to as univariate regression. On the other hand, two or more dependent variables constitute multivariate regression. Simple regression has only one predictor or target variable, while multivariate regression has more than one predictor variable.

Since linear regression in the base algorithm for all the different types of regression mentioned previously, in the next section, we will cover linear regression in detail.

LINEAR REGRESSION

The term "linear" refers to the linearity of parameters. Parameters are the coefficients of predictor variables in the linear regression equation. The following formula represents the linear regression equation:

$$y = \beta_0 + \beta_1 X + \epsilon$$

Figure 3.25: Formula for linear regression

Here, y is termed a dependent variable (output); it is continuous. X is an independent variable or feature (input). β0 and β1 are parameters. Є is the error component, which is the difference between the actual and predicted values of y. Since linear regression requires the variable to be linear, it is not used much in the real world. However, it is useful for high-level predictions, such as the sales revenue of a product given the price and advertising cost.

We can use the scikit-learn library to implement the linear regression algorithm in Python with the following code:

```
from sklearn.linear_model import LinearRegression
linreg = LinearRegression()
linreg.fit(X,y)
coefficient = linreg.coef_
intercept = linreg.intercept_
linreg.predict(X)
```

Here, we created the base model using the **LinearRegression** class of scikit-learn. Then, we trained the model using the **fit** function. Now that our linear regression model has been trained, we can use the **coef_** and **intercept_** parameters of the model to get the parameters and error components, as we discussed previously. Here, **X** represents a DataFrame of independent variables, whereas **y** represents a DataFrame of dependent variables. The trained model can then be used to make predictions using the **predict** function.

In the next section, we will solve an exercise to get a better understanding of regression analysis.

EXERCISE 3.06: REGRESSION ANALYSIS USING TEXTUAL DATA

In this exercise, we will make use of linear regression to predict the overall ratings from the reviews of musical instruments on Amazon. Follow these steps to implement this exercise:

> **NOTE**
>
> The dataset for this exercise can be downloaded from
> https://packt.live/3hQ6UEe.

1. Open a Jupyter Notebook.

2. Insert a new cell and add the following code to import the necessary packages:

```
import pandas as pd
import matplotlib.pyplot as plt
%matplotlib inline
import re
import string
from nltk import word_tokenize
from nltk.stem import WordNetLemmatizer
from sklearn.feature_extraction.text import TfidfVectorizer
from collections import Counter
from pylab import *
import nltk
nltk.download('punkt')
nltk.download('wordnet')
import warnings
warnings.filterwarnings('ignore')
```

3. Read the given data file in **JSON** format using **pandas**. Add the following code to implement this:

```
review_data = pd.read_json\
            ('data/reviews_Musical_Instruments_5.json', \
             lines=True)
review_data[['reviewText', 'overall']].head()
```

The preceding code generates the following output:

	reviewText	overall
0	Not much to write about here, but it does exac...	5
1	The product does exactly as it should and is q...	5
2	The primary job of this device is to block the...	5
3	Nice windscreen protects my MXL mic and preven...	5
4	This pop filter is great. It looks and perform...	5

Figure 3.26: Reviews of musical instruments stored as a DataFrame

4. Use a **lambda** function to extract tokens from each **'reviewText'** of this DataFrame, lemmatize them, and concatenate them side by side. Then, use the **join** function to concatenate a list of words into a single sentence. In order to replace anything other than alphabets, digits, and whitespaces with blank space, use the regular expression method (**re**). Add the following code to implement this:

```
lemmatizer = WordNetLemmatizer()
review_data['cleaned_review_text'] = review_data['reviewText']\
                                    .apply(lambda x : ' '.join\
                                    ([lemmatizer.lemmatize\
                                      (word.lower()) \
                                    for word in word_tokenize\
                                    (re.sub(r'([^\s\w]|_)+', ' ',\
                                    str(x)))]))
review_data[['cleaned_review_text', 'reviewText', 'overall']].head()
```

The preceding code generates the following output:

	cleaned_review_text	reviewText	overall
0	not much to write about here but it doe exactl...	Not much to write about here, but it does exac...	5
1	the product doe exactly a it should and is qui...	The product does exactly as it should and is q...	5
2	the primary job of this device is to block the...	The primary job of this device is to block the...	5
3	nice windscreen protects my mxl mic and preven...	Nice windscreen protects my MXL mic and preven...	5
4	this pop filter is great it look and performs ...	This pop filter is great. It looks and perform...	5

Figure 3.27: Review texts before and after cleaning, along with their overall scores

5. Create a **DataFrame** from the TFIDF matrix representation of cleaned **reviewText**. Add the following code to do this:

```
tfidf_model = TfidfVectorizer(max_features=500)
tfidf_df = pd.DataFrame(tfidf_model.fit_transform\
            (review_data['cleaned_review_text']).todense())
tfidf_df.columns = sorted(tfidf_model.vocabulary_)
tfidf_df.head()
```

The preceding code generates the following output:

	10	100	12	20	34	able	about	accurate	acoustic	actually	...	won	work	worked	worth	would	wrong	year	yet	you
0	0.0	0.0	0.0	0.0	0.0	0.000000	0.159684	0.0	0.0	0.0	...	0.0	0.134327	0.0	0.0	0.000000	0.0	0.0	0.0	0.000000
1	0.0	0.0	0.0	0.0	0.0	0.000000	0.000000	0.0	0.0	0.0	...	0.0	0.085436	0.0	0.0	0.000000	0.0	0.0	0.0	0.067074
2	0.0	0.0	0.0	0.0	0.0	0.000000	0.000000	0.0	0.0	0.0	...	0.0	0.000000	0.0	0.0	0.115312	0.0	0.0	0.0	0.079880
3	0.0	0.0	0.0	0.0	0.0	0.339573	0.000000	0.0	0.0	0.0	...	0.0	0.000000	0.0	0.0	0.000000	0.0	0.0	0.0	0.000000
4	0.0	0.0	0.0	0.0	0.0	0.000000	0.000000	0.0	0.0	0.0	...	0.0	0.000000	0.0	0.0	0.000000	0.0	0.0	0.0	0.303608

5 rows × 500 columns

Figure 3.28: TFIDF representation as a DataFrame

6. Use sklearn's **LinearRegression()** function to fit a linear regression model on this TFIDF DataFrame. Add the following code to do this:

```
from sklearn.linear_model import LinearRegression
linreg = LinearRegression()
linreg.fit(tfidf_df,review_data['overall'])
linreg.coef_
```

The preceding code generates the following output:

```
array([-1.93271993e-01,  5.65226131e-01,  5.63243687e-01, -1.84418658e-01,
       -6.32257431e-02,  3.05320627e-01,  4.95264614e-01,  5.21333693e-01,
        2.65736989e-01,  4.00058256e-01,  5.64020424e-01,  7.56022958e-01,
        1.00174846e-02, -3.06429115e-01, -3.12104234e-01,  3.38294736e-01,
       -6.05747380e-01, -1.04123996e-01,  5.58669738e-02, -1.13320890e-01,
        4.79471129e-01,  1.49528459e-01,  7.79094545e-01, -3.63399268e-01,
        1.25993539e-01, -6.29415062e-02,  4.94517373e-01, -3.34989132e-01,
        2.55374355e-01,  8.84676972e-02, -3.68013360e-01, -1.09910777e-01,
       -7.09777794e-03, -5.15547511e-02,  1.17415090e-01, -8.89213726e-02,
        1.06398798e+00, -1.19791236e+00, -1.14906460e+00,  1.55215016e-01,
       -5.05283241e-01,  2.43200389e-01,  8.56413437e-02, -3.74044994e-02,
       -7.31390217e-03,  9.63911076e-01, -7.82062558e-02,  1.50616236e-01,
       -9.35299622e-02,  1.87239631e-02,  9.34145997e-02,  1.18038260e+00,
       -3.79855115e-01,  4.51076351e-02,  1.11808544e-01,  7.22506502e-03,
        3.60057791e-01,  2.35459334e-01,  1.15359278e-01, -2.48993670e-01,
        1.34437898e-01, -2.99424905e-01, -1.00687767e-01, -3.10436924e-01,
        2.44420457e-02,  1.34593395e-01,  1.52613968e-01,  1.14304224e-01,
        8.46643557e-02, -9.06292369e-02,  1.88909690e-01,  1.71488133e-01,
       -1.37036225e+00,  3.67418288e-01,  3.00925842e-01,  3.45914386e-01,
       -1.39496654e-01,  6.68231981e-02,  5.38717132e-01,  6.48768917e-01,
```

Figure 3.29: Coefficients of the linear regression model

The preceding output shows the coefficients of the different features of the trained model.

Please note that *Figure 3.29* is truncated.

7. To check the intercept or the error term of the linear regression model, type the following code:

```
linreg.intercept_
```

The preceding code generates the following output:

```
4.218882428983381
```

8. To check the prediction in a TFIDF DataFrame, write the following code:

```
linreg.predict(tfidf_df)
```

The preceding code generates the following output:

```
array([4.19200071, 4.25771652, 4.23084868, …, 4.40384767,
        4.49036403, 4.14791976])
```

9. Finally, use this model to predict the **'overall'** score and store it in a column called **'predicted_score_from_linear_regression'**. Add the following code to implement this:

```
review_data['predicted_score_from_linear_regression'] = \
linreg.predict(tfidf_df)
review_data[['overall', \
            'predicted_score_from_linear_regression']].head(10)
```

The preceding code generates the following output:

	overall	predicted_score_from_linear_regression
0	5	4.192001
1	5	4.257717
2	5	4.230849
3	5	4.085927
4	5	4.851061
5	5	4.955069
6	5	4.446274
7	3	3.888593
8	5	4.941788
9	5	4.513824

Figure 3.30: Actual scores and predictions of the linear regression model

From the preceding table, we can see how the actual and predicted score varies for different instances. We will use this table later to evaluate the performance of the model.

> **NOTE**
>
> To access the source code for this specific section, please refer to https://packt.live/2P58eqy.
>
> You can also run this example online at https://packt.live/335pLqV.

You have just learned how to perform regression analysis on given data. In the next section, you will learn about tree methods.

TREE METHODS

There are several algorithms that have both classification and regression forms. Tree-based methods are instances of such cases. In the context of machine learning, "tree" refers to a structure that aids decision-making—hence, the term **decision tree**. Tree-based methods have high accuracy and unlike linear methods, they model non-linear relationships as well. Additionally, decision trees handle categorical variables much better than linear regression.

Let us use the example of a hotel trying to identify if the reviews provided by their patrons have a positive sentiment or a negative one. So, the reviews needed to be classified into two classes, namely, positive sentiments and negative sentiments. A data scientist working for the hotel can create a dataset of all online reviews of their hotel and create a decision tree, as shown in the following diagram:

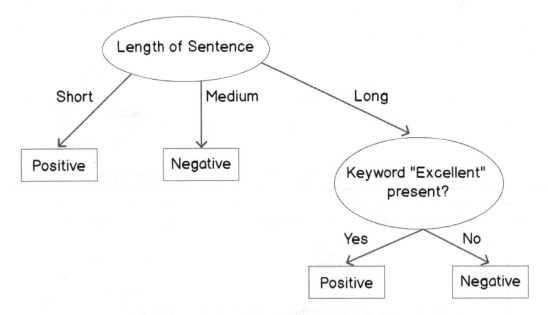

Figure 3.31: Decision tree

In the preceding diagram, the first decision is made based on the length of the sentences. He finds that the short length reviews generally have a positive sentiment, whereas a medium length review has a negative sentiment. For reviews that were longer, he had to rely on keywords to determine the sentiment as longer length reviews are almost equally likely to be positive or negative. If the *excellent* keyword is present, the review belongs to the **positive** sentiment; otherwise, it belongs to the **negative** sentiment.

We can make use of the scikit-learn library to implement the decision tree algorithm in Python using the following code:

```
from sklearn import tree
dtc = tree.DecisionTreeClassifier()
dtc = dtc.fit(X, y)
predicted_labels = dtc.predict(X)
```

Here, we created the base model using the **DecisionTreeClassifier** class of scikit-learn. Then, we trained the model using the **fit** function. The trained model can then be used to make predictions using the **predict** function. Here, **X** represents a DataFrame of independent variables, whereas **y** represents a DataFrame of dependent variables.

EXERCISE 3.07: TREE-BASED METHODS – DECISION TREE

In this exercise, we will use the tree-based method known as decision trees to predict the overall scores and labels of reviews of patio, lawn, and garden products on Amazon. Follow these steps to implement this exercise:

> **NOTE**
>
> To download the dataset for this exercise, visit https://packt.live/3gczb7P.

1. Open a Jupyter Notebook.

2. Insert a new cell and add the following code to import the necessary packages:

```
import pandas as pd
import matplotlib.pyplot as plt
%matplotlib inline
import re
import string
from nltk import word_tokenize
from nltk.stem import WordNetLemmatizer
from sklearn.feature_extraction.text import TfidfVectorizer
from collections import Counter
from pylab import *
```

```
import nltk
nltk.download('wordnet')
nltk.download('punkt')
import warnings
warnings.filterwarnings('ignore')
```

3. Now, read the given data file in JSON format using pandas. Add the following code to implement this:

```
data_patio_lawn_garden = pd.read_json\
                            ('data/'\
                            'reviews_Patio_Lawn_and_Garden_5.json',\
                            lines = True)
data_patio_lawn_garden[['reviewText', 'overall']].head()
```

The preceding code generates the following output:

	reviewText	overall
0	Good USA company that stands behind their prod...	4
1	This is a high quality 8 ply hose. I have had ...	5
2	It's probably one of the best hoses I've ever ...	4
3	I probably should have bought something a bit ...	5
4	I bought three of these 5/8-inch Flexogen hose...	5

Figure 3.32: Storing reviews as a DataFrame

4. Use the **lambda** function to extract tokens from each **'reviewText'** of this DataFrame, lemmatize them using **WorrdNetLemmatizer**, and concatenate them side by side. Use the **join** function to concatenate a list of words into a single sentence. Use the regular expression method (**re**) to replace anything other than letters, digits, and whitespaces with blank spaces. Add the following code to do this:

```
lemmatizer = WordNetLemmatizer()
data_patio_lawn_garden['cleaned_review_text'] = \
data_patio_lawn_garden['reviewText']\
.apply(lambda x : ' '.join([lemmatizer.lemmatize(word.lower()) \
        for word in word_tokenize(re.sub(r'([^\s\w]|_)+', ' ', \
        str(x)))]))
```

```
data_patio_lawn_garden[['cleaned_review_text', 'reviewText',\
                        'overall']].head()
```

The preceding code generates the following output:

	cleaned_review_text	reviewText	overall
0	good usa company that stand behind their produ...	Good USA company that stands behind their prod...	4
1	this is a high quality 8 ply hose i have had g...	This is a high quality 8 ply hose. I have had ...	5
2	it s probably one of the best hose i ve ever h...	It's probably one of the best hoses I've ever ...	4
3	i probably should have bought something a bit ...	I probably should have bought something a bit ...	5
4	i bought three of these 5 8 inch flexogen hose...	I bought three of these 5/8-inch Flexogen hose...	5

Figure 3.33: Review text before and after cleaning, along with overall scores

5. Create a DataFrame from the TFIDF matrix representation of the cleaned version of **reviewText** with the following code:

```
tfidf_model = TfidfVectorizer(max_features=500)
tfidf_df = pd.DataFrame(tfidf_model.fit_transform\
            (data_patio_lawn_garden['cleaned_review_text'])\
            .todense())
tfidf_df.columns = sorted(tfidf_model.vocabulary_)
tfidf_df.head()
```

The preceding code generates the following output:

	10	20	34	8217	able	about	actually	add	after	again	...	work	worked	working	worth	would	yard	year	yet	you
0	0.0	0.0	0.0	0.0	0.0	0.000000	0.0	0.0	0.120568	0.0	...	0.0	0.0	0.0	0.0	0.0	0.000000	0.0	0.000000	0.161561
1	0.0	0.0	0.0	0.0	0.0	0.000000	0.0	0.0	0.000000	0.0	...	0.0	0.0	0.0	0.0	0.0	0.000000	0.0	0.000000	0.000000
2	0.0	0.0	0.0	0.0	0.0	0.000000	0.0	0.0	0.000000	0.0	...	0.0	0.0	0.0	0.0	0.0	0.116566	0.0	0.216988	0.000000
3	0.0	0.0	0.0	0.0	0.0	0.000000	0.0	0.0	0.000000	0.0	...	0.0	0.0	0.0	0.0	0.0	0.000000	0.0	0.000000	0.000000
4	0.0	0.0	0.0	0.0	0.0	0.064347	0.0	0.0	0.070857	0.0	...	0.0	0.0	0.0	0.0	0.0	0.083019	0.0	0.000000	0.000000

5 rows × 500 columns

Figure 3.34: TFIDF representation as a DataFrame

6. The following lines of code are used to create a new column called target, which will have 0 if the **'overall'** parameter is less than 4; otherwise, it will have 1:

```
data_patio_lawn_garden['target'] = data_patio_lawn_garden\
                                    ['overall'].apply\
                                    (lambda x : 0 if x<=4 else 1)
data_patio_lawn_garden['target'].value_counts()
```

The preceding code generates the following output:

```
1      7037
0      6235
Name:  target, dtype: int64
```

7. Use sklearn's **tree()** function to fit a decision tree classification model on the TFIDF DataFrame we created earlier. Add the following code to do this:

```
from sklearn import tree
dtc = tree.DecisionTreeClassifier()
dtc = dtc.fit(tfidf_df, data_patio_lawn_garden['target'])
data_patio_lawn_garden['predicted_labels_dtc'] = dtc.predict\
                                                 (tfidf_df)
```

8. Use pandas' **crosstab** function to compare the results of the classification model with the actual classes (**'target'**, in this case) of the reviews. Add the following code to do this:

```
pd.crosstab(data_patio_lawn_garden['target'], \
            data_patio_lawn_garden['predicted_labels_dtc'])
```

The preceding code generates the following output:

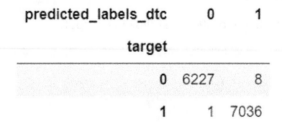

predicted_labels_dtc	0	1
target		
0	6227	8
1	1	7036

Figure 3.35: Crosstab between actual target values and predicted labels

Here, we can see **6627** instances with a target label of **0** correctly classified, and **8** such instances wrongly classified. Furthermore, **7036** instances with a target label of **1** are correctly classified, whereas **1** such instance is wrongly classified.

9. Use sklearn's **tree()** function to fit a decision tree regression model on the TFIDF representation of these reviews after cleaning. To predict the overall scores using this model, add the following code:

```
from sklearn import tree
dtr = tree.DecisionTreeRegressor()
dtr = dtr.fit(tfidf_df, data_patio_lawn_garden['overall'])
data_patio_lawn_garden['predicted_values_dtr'] = dtr.predict\
                                          (tfidf_df)
data_patio_lawn_garden[['predicted_values_dtr', \
                    'overall']].head(10)
```

The preceding code generates the following output:

	predicted_values_dtr	overall
0	4.0	4
1	5.0	5
2	4.0	4
3	5.0	5
4	5.0	5
5	5.0	5
6	5.0	5
7	5.0	5
8	5.0	5
9	4.0	4

Figure 3.36: Overall scores with predicted values

From the preceding table, we can see how the actual and predicted scores vary for different instances. We will use this table later to evaluate the performance of the model.

> **NOTE**
>
> To access the source code for this specific section, please refer to https://packt.live/39CHhEc.
>
> You can also run this example online at https://packt.live/39EwKZ6.

Next, we will look at another tree-based method, random forest.

RANDOM FOREST

Imagine that you must decide whether to join a university. In one scenario, you ask only one person about the quality of the education the university provides. In another scenario, you ask several career counselors and academicians about this. Which scenario do you think would help you make a better and the most stable decision? The second one, right? This is because, in the first case, the only person you are consulting may be biased. "Wisdom of the crowd" tends to remove biases, thereby aiding better decision-making.

In general terms, a forest is a collection of different types of trees. The same definition holds true in the case of machine learning as well. Instead of using a single decision tree for prediction, we use several of them.

In the scenario we described earlier, the first case is equivalent to using a single decision tree, whereas the second one is equivalent to using several—that is, using a forest. In a random forest, an individual tree's vote impacts the final decision. Just like decision trees, random forest is capable of carrying out both classification and regression tasks.

An advantage of the random forest algorithm is that it uses a sampling technique called bagging, which prevents **overfitting**. Bagging is the process of training meta-algorithms on a different subsample of the data and then combining these to create a better model. Overfitting refers to cases where a model learns the training dataset so well that it is unable to generalize or perform well on another validation/test dataset.

Random forests also aid in understanding the importance of predictor variables and features. However, building a random forest often takes a huge amount of time and memory. We can make use of the scikit-learn library to implement the random forest algorithm in Python using the following code:

```
from sklearn.ensemble import RandomForestClassifier
rfc = RandomForestClassifier()
rfc = rfc.fit(X, y)
predicted_labels = rfc.predict(X)
```

Here, we created the base model using the **RandomForestClassifier** class of scikit-learn. Then, we trained the model using the **fit** function. The trained model can then be used to make predictions using the **predict** function. **X** represents a DataFrame of independent variables, whereas **y** represents a DataFrame of dependent variables.

GRADIENT BOOSTING MACHINE AND EXTREME GRADIENT BOOST

There are various other tree-based algorithms, such as **gradient boosting machines** (**GBM**) and **extreme gradient boosting** (**XGBoost**). Boosting works by combining rough, less complex, or "weak" models into a single prediction that is more accurate than any single model. Iteratively, a subset of the training dataset is ingested into a "weak" algorithm or simple algorithm (such as a decision tree) to generate a weak model. These weak models are then combined to form the final prediction.

GBM makes use of classification trees as the weak algorithm. The results are generated by improving estimations from these weak models using a differentiable loss function, which gives us the performance of the model by calculating how far the prediction is from the actual value. The model fits consecutive trees by considering the net loss of the previous trees; therefore, each tree is partially present in the final solution.

XGBoost is an enhanced version of GBM that is portable and distributed, which means that it can easily be used in different architectures and can use multiple cores (a single machine) or multiple machines (clusters). As a bonus, the **XGBoost** library is written in C++, which makes it fast. It is also useful when working with a huge dataset as it allows you to store data on an external disk rather than load all the data into memory. The main reasons for the popularity of XGBoost are as follows:

- Ability to automatically deal with missing values

- High-speed execution

- High accuracy, if properly trained

- Support for distributed frameworks such as Hadoop and Spark

XGBoost uses a weighted combination of weak learners during the training phase.

We can make use of the **xgboost** library to implement the XGBoost algorithm in Python using the following code:

```
from xgboost import XGBClassifier
xgb_clf=XGBClassifier()
xgb_clf = xgb_clf.fit(X, y)
predicted_labels = rfc.predict(X)
```

Here, we created the base model using the **XGBClassifier** class of **xgboost**. Then, we trained the model using the **fit** function. The trained model can then be used to make predictions using the predict function. Here, **X** represents a DataFrame of independent variables, whereas **y** represents a DataFrame of dependent variables. To get the important features for the trained model, we can use the following code:

```
pd.DataFrame({'word':X.columns,'importance':xgb_clf.feature_
importances_})
```

Let's perform some exercises to get a better understanding of tree-based methods.

EXERCISE 3.08: TREE-BASED METHODS — RANDOM FOREST

In this exercise, we will use the tree-based method random forest to predict the overall scores and labels of reviews of patio, lawn, and garden products on Amazon. Follow these steps to implement this exercise:

> **NOTE**
>
> To download the dataset for this exercise, visit https://packt.live/3gczb7P.

1. Open a Jupyter Notebook. Insert a new cell and add the following code to import the necessary packages:

```
import pandas as pd
import matplotlib.pyplot as plt
%matplotlib inline
import re
import string
from nltk import word_tokenize
from nltk.stem import WordNetLemmatizer
from sklearn.feature_extraction.text import TfidfVectorizer
from collections import Counter
from pylab import *
import nltk
nltk.download('punkt')
nltk.download('wordnet')
import warnings
warnings.filterwarnings('ignore')
```

2. Now, read the given data file in **JSON** format using **pandas**. Add the following code to implement this:

```
data_patio_lawn_garden = pd.read_json\
                         ('data/'\
                          'reviews_Patio_Lawn_and_Garden_5.json',\
                          lines = True)
data_patio_lawn_garden[['reviewText', 'overall']].head()
```

The preceding code generates the following output:

	reviewText	overall
0	Good USA company that stands behind their prod...	4
1	This is a high quality 8 ply hose. I have had ...	5
2	It's probably one of the best hoses I've ever ...	4
3	I probably should have bought something a bit ...	5
4	I bought three of these 5/8-inch Flexogen hose...	5

Figure 3.37: Storing reviews as a DataFrame

3. Use a **lambda** function to extract tokens from each **reviewText** of this DataFrame, lemmatize them using **WordNetLemmatizer**, and concatenate them side by side. Use the **join** function to concatenate a list of words into a single sentence. Use a regular expression (**re**) to replace anything other than letters, digits, and whitespaces with blank spaces. Add the following code to do this:

```
lemmatizer = WordNetLemmatizer()
data_patio_lawn_garden['cleaned_review_text'] = \
data_patio_lawn_garden['reviewText']\
.apply(lambda x : ' '.join([lemmatizer.lemmatize(word.lower()) \
        for word in word_tokenize(re.sub(r'([^\s\w]|_)+', ' ', \
                            str(x)))]))
data_patio_lawn_garden[['cleaned_review_text', 'reviewText', \
                    'overall']].head()
```

The preceding code generates the following output:

	cleaned_review_text	reviewText	overall
0	good usa company that stand behind their produ...	Good USA company that stands behind their prod...	4
1	this is a high quality 8 ply hose i have had g...	This is a high quality 8 ply hose. I have had ...	5
2	it s probably one of the best hose i ve ever h...	It's probably one of the best hoses I've ever ...	4
3	i probably should have bought something a bit ...	I probably should have bought something a bit ...	5
4	i bought three of these 5 8 inch flexogen hose...	I bought three of these 5/8-inch Flexogen hose...	5

Figure 3.38: Review text before and after cleaning, along with overall scores

4. Create a **DataFrame** from the TFIDF matrix representation of the cleaned version of **reviewText** with the following code:

```
tfidf_model = TfidfVectorizer(max_features=500)
tfidf_df = pd.DataFrame(tfidf_model.fit_transform\
            (data_patio_lawn_garden['cleaned_review_text'])\
            .todense())
tfidf_df.columns = sorted(tfidf_model.vocabulary_)
tfidf_df.head()
```

The preceding code generates the following output:

	10	20	34	8217	able	about	actually	add	after	again	...	work	worked	working	worth	would	yard	year	yet	you
0	0.0	0.0	0.0	0.0	0.0	0.000000	0.0	0.0	0.120568	0.0	...	0.0	0.0	0.0	0.0	0.0	0.000000	0.0	0.000000	0.161561
1	0.0	0.0	0.0	0.0	0.0	0.000000	0.0	0.0	0.000000	0.0	...	0.0	0.0	0.0	0.0	0.0	0.000000	0.0	0.000000	0.000000
2	0.0	0.0	0.0	0.0	0.0	0.000000	0.0	0.0	0.000000	0.0	...	0.0	0.0	0.0	0.0	0.0	0.116566	0.0	0.216988	0.000000
3	0.0	0.0	0.0	0.0	0.0	0.000000	0.0	0.0	0.000000	0.0	...	0.0	0.0	0.0	0.0	0.0	0.000000	0.0	0.000000	0.000000
4	0.0	0.0	0.0	0.0	0.0	0.064347	0.0	0.0	0.070857	0.0	...	0.0	0.0	0.0	0.0	0.0	0.083019	0.0	0.000000	0.000000

5 rows × 500 columns

Figure 3.39: TFIDF representation as a DataFrame

5. Add the following lines of code to create a new column called target, which will have 0 if the **overall** parameter is less than 4; otherwise, it will have 1:

```
data_patio_lawn_garden['target'] = data_patio_lawn_garden\
                                ['overall'].apply\
                                (lambda x : 0 if x<=4 else 1)
data_patio_lawn_garden['target'].value_counts()
```

The preceding code generates the following output:

```
1       7037
0       6235
Name:   target, dtype: int64
```

6. Now, define a generic function for all the classifier models. Add the following code to do this:

```
def clf_model(model_type, X_train, y):
    model = model_type.fit(X_train,y)
    predicted_labels = model.predict(tfidf_df)
    return predicted_labels
```

7. Train three kinds of classifier models—namely, random forest, gradient boosting machines, and XGBoost. For random forest, we predict the class labels of the given set of review texts and compare them with their actual class—that is, the target, using crosstabs. Add the following code to do this:

```
from sklearn.ensemble import RandomForestClassifier
rfc = RandomForestClassifier(n_estimators=20,max_depth=4,\
                        max_features='sqrt',random_state=1)
data_patio_lawn_garden['predicted_labels_rfc'] = \
clf_model(rfc, tfidf_df, data_patio_lawn_garden['target'])
pd.crosstab(data_patio_lawn_garden['target'], \
        data_patio_lawn_garden['predicted_labels_rfc'])
```

The preceding code generates the following output:

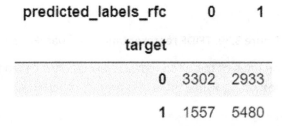

predicted_labels_rfc	0	1
target		
0	3302	2933
1	1557	5480

Figure 3.40: Crosstab between actual target values and predicted labels

Here, we can see **3302** instances with a target label of 0 correctly classified, and **2933** such instances wrongly classified. Furthermore, **5480** instances with a target label of 1 are correctly classified, whereas **1557** such instances are wrongly classified.

8. Now, define a generic function for all regression models. Add the following code to do this:

```
def reg_model(model_type, X_train, y):
    model = model_type.fit(X_train,y)
    predicted_values = model.predict(tfidf_df)
    return predicted_values
```

9. Train three kinds of regression models: random forest, gradient boosting machines, and XGBoost. For random forest, we predict the overall score of the given set of review texts. Add the following code to do this:

```
from sklearn.ensemble import RandomForestRegressor
rfg = RandomForestRegressor(n_estimators=20,max_depth=4,\
                        max_features='sqrt',random_state=1)
data_patio_lawn_garden['predicted_values_rfg'] = \
reg_model(rfg, tfidf_df, data_patio_lawn_garden['overall'])
data_patio_lawn_garden[['overall', \
                    'predicted_values_rfg']].head(10)
```

The preceding code generates the following output:

	overall	predicted_values_rfg
0	4	4.236717
1	5	4.341767
2	4	4.219413
3	5	4.134852
4	5	4.147218
5	5	4.252751
6	5	4.190971
7	5	4.251688
8	5	4.251610
9	4	4.262498

Figure 3.41: Actual overall score and predicted values using a random forest regressor

From the preceding table, we can see how the actual and predicted scores vary for different instances. We will use this table later to evaluate the performance of the model.

> **NOTE**
>
> To access the source code for this specific section, please refer to https://packt.live/33aowa4.
>
> You can also run this example online at https://packt.live/2P8a89V.

Now, let's perform a similar task using the XGBoost method.

EXERCISE 3.09: TREE-BASED METHODS – XGBOOST

In this exercise, we will use the tree-based method XGBoost to predict the overall scores and labels of reviews of patio, lawn, and garden products on Amazon.

> **NOTE**
>
> To download the dataset for this exercise, visit https://packt.live/3gczb7P.

Follow these steps to implement this exercise:

1. Open a Jupyter Notebook.

2. Insert a new cell and add the following code to import the necessary packages:

```
import pandas as pd
import matplotlib.pyplot as plt
%matplotlib inline
import re
import string
from nltk import word_tokenize
from nltk.stem import WordNetLemmatizer
from sklearn.feature_extraction.text import TfidfVectorizer
from collections import Counter
from pylab import *
```

```
import nltk
nltk.download('punkt')
nltk.download('wordnet')
import warnings
warnings.filterwarnings('ignore')
```

3. Now, read the given data file in **JSON** format using **pandas**. Add the following code to implement this:

```
data_patio_lawn_garden = pd.read_json\
                         ('data/'\
                         'reviews_Patio_Lawn_and_Garden_5.json',\
                         lines = True)
data_patio_lawn_garden[['reviewText', 'overall']].head()
```

The preceding code generates the following output:

	reviewText	overall
0	Good USA company that stands behind their prod...	4
1	This is a high quality 8 ply hose. I have had ...	5
2	It's probably one of the best hoses I've ever ...	4
3	I probably should have bought something a bit ...	5
4	I bought three of these 5/8-inch Flexogen hose...	5

Figure 3.42: Storing reviews as a DataFrame

4. Use a **lambda** function to extract tokens from each **'reviewText'** of this DataFrame, lemmatize them using **WorrdNetLemmatizer**, and concatenate them side by side. Use the **join** function to concatenate a list of words into a single sentence. Use the regular expression method (**re**) to replace anything other than letters, digits, and whitespaces with blank spaces. Add the following code to do this:

```
lemmatizer = WordNetLemmatizer()
data_patio_lawn_garden['cleaned_review_text'] = \
data_patio_lawn_garden['reviewText'].apply(lambda x : ' '.join\
                                ([lemmatizer.lemmatize\
                                (word.lower()) \
                                for word in word_tokenize\
                                (re.sub\
```

```
                                              (r'([^\s\w]|_)+', ' ', \
                                      str(x)))]))
     data_patio_lawn_garden[['cleaned_review_text', 'reviewText', \
                             'overall']].head()
```

The preceding code generates the following output:

	cleaned_review_text	reviewText	overall
0	good usa company that stand behind their produ...	Good USA company that stands behind their prod...	4
1	this is a high quality 8 ply hose i have had g...	This is a high quality 8 ply hose. I have had ...	5
2	it s probably one of the best hose i ve ever h...	It's probably one of the best hoses I've ever ...	4
3	i probably should have bought something a bit ...	I probably should have bought something a bit ...	5
4	i bought three of these 5 8 inch flexogen hose...	I bought three of these 5/8-inch Flexogen hose...	5

Figure 3.43: Review text before and after cleaning, along with overall scores

5. Create a DataFrame from the TFIDF matrix representation of the cleaned version of **reviewText** with the following code:

```
tfidf_model = TfidfVectorizer(max_features=500)
tfidf_df = pd.DataFrame(tfidf_model.fit_transform\
                (data_patio_lawn_garden['cleaned_review_text'])\
                .todense())
tfidf_df.columns = sorted(tfidf_model.vocabulary_)
tfidf_df.head()
```

The preceding code generates the following output:

	10	20	34	8217	able	about	actually	add	after	again	...	work	worked	working	worth	would	yard	year	yet	you
0	0.0	0.0	0.0	0.0	0.0	0.000000	0.0	0.0	0.120568	0.0	...	0.0	0.0	0.0	0.0	0.0	0.000000	0.0	0.000000	0.161561
1	0.0	0.0	0.0	0.0	0.0	0.000000	0.0	0.0	0.000000	0.0	...	0.0	0.0	0.0	0.0	0.0	0.000000	0.0	0.000000	0.000000
2	0.0	0.0	0.0	0.0	0.0	0.000000	0.0	0.0	0.000000	0.0	...	0.0	0.0	0.0	0.0	0.0	0.116566	0.0	0.216988	0.000000
3	0.0	0.0	0.0	0.0	0.0	0.000000	0.0	0.0	0.000000	0.0	...	0.0	0.0	0.0	0.0	0.0	0.000000	0.0	0.000000	0.000000
4	0.0	0.0	0.0	0.0	0.0	0.064347	0.0	0.0	0.070857	0.0	...	0.0	0.0	0.0	0.0	0.0	0.083019	0.0	0.000000	0.000000

5 rows × 500 columns

Figure 3.44: TFIDF representation as a DataFrame

6. The following lines of code are used to create a new column called target, which will have 0 if the **'overall'** parameter is less than 4; otherwise, it will have 1:

```
data_patio_lawn_garden['target'] = data_patio_lawn_garden\
                                   ['overall'].apply\
                                   (lambda x : 0 if x<=4 else 1)
data_patio_lawn_garden['target'].value_counts()
```

The preceding code generates the following output:

```
1       7037
0       6235
Name:   target, dtype: int64
```

7. Now, define a generic function for all the classifier models. Add the following code to do this:

```
def clf_model(model_type, X_train, y):
    model = model_type.fit(X_train,y)
    predicted_labels = model.predict(tfidf_df)
    return predicted_labels
```

8. Predict the class labels of the given set of **reviewText** and compare it with their actual class, that is, the target, using the **crosstab** function. Add the following code to do this:

```
pip install xgboost
from xgboost import XGBClassifier
xgb_clf=XGBClassifier(n_estimators=20,learning_rate=0.03,\
                      max_depth=5,subsample=0.6,\
                      colsample_bytree= 0.6,reg_alpha= 10,\
                      seed=42)
data_patio_lawn_garden['predicted_labels_xgbc'] = \
clf_model(xgb_clf, tfidf_df, data_patio_lawn_garden['target'])
pd.crosstab(data_patio_lawn_garden['target'], \
            data_patio_lawn_garden['predicted_labels_xgbc'])
```

The preceding code generates the following output:

predicted_labels_xgbc	0	1
target		
0	4300	1935
1	2164	4873

Figure 3.45: Crosstab between actual target values and predicted labels using XGBoost

Here, we can see **4300** instances with a target label of 0 correctly classified, and **1935** such instances wrongly classified. Furthermore, **2164** instances with a target label of 1 are correctly classified, whereas **4873** such instances are wrongly classified.

9. Now, define a generic function for all the regression models. Add the following code to do this:

```
def reg_model(model_type, X_train, y):
    model = model_type.fit(X_train,y)
    predicted_values = model.predict(tfidf_df)
    return predicted_values
```

10. Predict the overall score of the given set of **reviewText**. Add the following code to do this:

```
from xgboost import XGBRegressor
xgbr = XGBRegressor(n_estimators=20,learning_rate=0.03,\
                    max_depth=5,subsample=0.6,\
                    colsample_bytree= 0.6,reg_alpha= 10,seed=42)
data_patio_lawn_garden['predicted_values_xgbr'] = \
reg_model(xgbr, tfidf_df, data_patio_lawn_garden['overall'])
data_patio_lawn_garden[['overall', \
                    'predicted_values_xgbr']].head(2)
```

The preceding code generates the following output:

	overall	predicted_values_xgbr
0	4	2.169383
1	5	2.210736

Figure 3.46: Actual overall score and predicted values using an XGBoost regressor

From the preceding table, we can see how the actual and predicted scores vary for different instances. We will use this table later to evaluate the performance of the model. With that, you have learned how to use tree-based methods to predict scores in data.

> **NOTE**
>
> To access the source code for this specific section, please refer to https://packt.live/2P5woBi.
>
> You can also run this example online at https://packt.live/2DfTa71.

In the next section, you will learn about sampling.

SAMPLING

Sampling is the process of creating a subset from a given set of instances. If you have 1,000 sentences in an article, out of which you choose 100 sentences for analysis, the subset of 100 sentences will be called a sample of the original article. This process is referred to as sampling.

Sampling is necessary when creating models for imbalanced datasets. For example, consider that the number of bad comments on a review board for a company is 10 and the number of good comments is 1,000. If we input this data as it is into the model, it will not give us accurate results; classifying all the comments as "good" will provide a near-perfect accuracy, which isn't really applicable to most real datasets. Thus, we need to reduce the number of good reviews to a smaller number before using it as input for training. There are different kinds of sampling methods, such as the following:

- **Simple random sampling**

  In this process, each instance of the set has an equal probability of being selected. For example, you have 10 balls of 10 different colors in a box. You need to select 4 out of 10 balls without looking at their color. In this case, each ball is equally likely to be selected. This is an instance of simple random sampling.

- **Stratified sampling**

  In this type of sampling, the original set is divided into parts called "strata", based on given criteria. Random samples are chosen from each of these "strata." For example, you have 100 sentences, out of which 80 are non-sarcastic and 20 are sarcastic. To extract a stratified sample of 10 sentences, you need to select 8 from 80 non-sarcastic sentences and 2 from 20 sarcastic sentences. This will ensure that the ratio of non-sarcastic to sarcastic sentences, that is, 80:20, remains unaltered in the sample that's selected.

- **Multi-Stage Sampling**

  If you are analyzing the social media posts of all the people in a country related to the current weather, the text data will be huge as it will consist of the weather conditions of different cities. Drawing a stratified sample would be difficult. In this case, it is recommended to first extract a stratified sample by region and then further sample it within regions, that is, by cities. This is basically performing stratified sampling at each and every stage.

To better understand these, let's perform a simple exercise.

EXERCISE 3.10: SAMPLING (SIMPLE RANDOM, STRATIFIED, AND MULTI-STAGE)

In this exercise, we will extract samples from an online retail dataset that contains details about the transactions of an e-commerce website with the help of simple random sampling, stratified sampling, and multi-stage sampling.

> **NOTE**
>
> To download the dataset for this exercise, visit https://packt.live/3fdsZuL.

Follow these steps to implement this exercise:

1. Open a Jupyter Notebook.

2. Insert a new cell and add the following code to import pandas and read the dataset:

```
!pip install xlrd
import pandas as pd
data = pd.read_excel('data/Online Retail.xlsx')
data.shape
```

The preceding code generates the following output:

```
(54190, 8)
```

3. We use pandas' **sample** function to extract a sample from the **DataFrame**. Add the following code to do this:

```
# selecting 10% of the data randomly
data_sample_random = data.sample(frac=0.1,random_state=42)
data_sample_random.shape
```

The preceding code generates the following output:

```
(54191, 8)
```

4. Use sklearn's **train_test_split** function to create stratified samples. Add the following code to do this:

```
from sklearn.model_selection import train_test_split
X_train, X_valid, y_train, y_valid = train_test_split\
                            (data, data['Country'],\
                             test_size=0.2, \
                             random_state=42,\
                             stratify = data['Country'])
```

You can confirm the stratified split by checking the percentage of each category in the country column after the split. To get the train set percentages, use the following code:

```
y_train.value_counts()/y_train.shape[0]
```

The following is part of the output of the preceding code:

```
United Kingdom        0.914319
Germany               0.017521
France                0.015789
EIRE                  0.015125
Spain                 0.004673
Netherlands           0.004376
Belgium               0.003818
Switzerland           0.003695
Portugal              0.002803
Australia             0.002323
Norway                0.002004
Italy                 0.001481
Channel Islands       0.001398
Finland               0.001283
Cyprus                0.001149
Sweden                0.000853
Unspecified           0.000823
Austria               0.000740
Denmark               0.000717
Japan                 0.000660
Poland                0.000630
```

Figure 3.47: The percentage of countries for the training set

5. Similarly, for the validation set, add the following code:

```
y_valid.value_counts()/y_valid.shape[0]
```

The following is part of the output of the preceding code:

```
United Kingdom        0.914322
Germany               0.017521
France                0.015796
EIRE                  0.015122
Spain                 0.004678
Netherlands           0.004373
Belgium               0.003820
Switzerland           0.003691
Portugal              0.002805
Australia             0.002325
Norway                0.002002
Italy                 0.001485
Channel Islands       0.001402
Finland               0.001283
Cyprus                0.001144
Sweden                0.000849
Unspecified           0.000821
Austria               0.000738
Denmark               0.000720
Japan                 0.000664
Poland                0.000627
```

Figure 3.48: The percentage of countries for the validation set

As we can see, the percentages of all countries are similar in both the train and validation sets.

6. We filter out the data in various stages and extract random samples from it. We will extract a random sample of 2% from those transactions by country that occurred in the United Kingdom, Germany, or France and where the corresponding quantity is greater than or equal to 2. Add the following code to implement this:

```
data_ugf = data[data['Country'].isin(['United Kingdom', \
                                      'Germany', 'France'])]
data_ugf_q2 = data_ugf[data_ugf['Quantity']>=2]
data_ugf_q2_sample = data_ugf_q2.sample(frac = .02, \
                                        random_state=42)
data_ugf_q2.shape
```

The preceding code generates the following output:

```
(356940, 8)
```

Now, add the following line of code:

```
data_ugf_q2_sample.shape
```

This will generate the following output:

```
(7139, 8)
```

We can see the reduction in size of the data when the filtering criteria is applied and then the reduction in size when a sample of the filtered data is taken. In this exercise, you learned about the three major sampling techniques that will help you create a good training dataset for the text classifier that you will learn how to build in the next section.

> **NOTE**
>
> To access the source code for this specific section, please refer to https://packt.live/2P7M4nD.
>
> You can also run this example online at https://packt.live/3jT8XsZ.

DEVELOPING A TEXT CLASSIFIER

A text classifier is a machine learning model that is capable of labeling texts based on their content. For instance, a text classifier will help you understand whether a random text statement is sarcastic or not. Presently, text classifiers are gaining importance as manually classifying huge amounts of text data is impossible. In the next few sections, we will learn about the different parts of text classifiers and implement them in Python.

FEATURE EXTRACTION

When dealing with text data, features denote its different attributes. Generally, they are numeric representations of the text. As we discussed in *Chapter 2, Feature Extraction Methods*, TFIDF representations of texts are one of the most popular ways of extracting features from them.

FEATURE ENGINEERING

Feature engineering is the art of extracting new features from existing ones. Extracting novel features, which tend to capture variation in data better, requires sound domain expertise.

REMOVING CORRELATED FEATURES

Correlation refers to the statistical relationship between two variables. Two highly correlated variables provide the same kind of information. For example, the remaining battery life of a laptop and its screen time are highly correlated. The battery life will decrease as the screen time increases. Regression models, including logistic regression, are unable to perform well when correlation between features exists. Thus, features with correlation beyond a certain threshold need to be removed. The most widely used correlation statistic is Pearson correlation, which can be calculated as follows:

$$corr(X, Y) = \frac{cov(X, Y)}{\sigma_x \sigma_y}$$

Figure 3.49: Pearson correlation

Here, *cov* is the covariance, σ is the standard deviation, and X and Y are two variables/ features of the training data that we are testing for correlation.

EXERCISE 3.11: REMOVING HIGHLY CORRELATED FEATURES (TOKENS)

In this exercise, we will remove highly correlated words from a TFIDF matrix representation of sklearn's **fetch_20newgroups** text dataset. Follow these steps to implement this exercise:

1. Open a Jupyter Notebook.

2. Insert a new cell and add the following code to import the necessary packages:

```
from sklearn.datasets import fetch_20newsgroups
import matplotlib as mpl
import pandas as pd
import numpy as np
import matplotlib.pyplot as plt
%matplotlib inline
import re
import string
from nltk import word_tokenize
from nltk.corpus import stopwords
from nltk.stem import WordNetLemmatizer
from sklearn.feature_extraction.text import TfidfVectorizer
from collections import Counter
from pylab import *
import nltk
nltk.download('stopwords')
nltk.download('punkt')
nltk.download('wordnet')
import warnings
warnings.filterwarnings('ignore')
```

3. We will be using stop words from the English language only. WordNet is the **lemmatizer** we will be using. Add the following code to implement this:

```
stop_words = stopwords.words('english')
stop_words = stop_words + list(string.printable)
lemmatizer = WordNetLemmatizer()
```

4. To specify the categories of news articles you want to fetch, add the following code:

```
categories= ['misc.forsale', 'sci.electronics', \
             'talk.religion.misc']
```

5. To fetch sklearn's **20newsgroups** text dataset, corresponding to the categories mentioned earlier, use the following lines of code:

```
news_data = fetch_20newsgroups(subset='train', \
                               categories=categories, \
                               shuffle=True, random_state=42, \
                               download_if_missing=True)
news_data_df = pd.DataFrame({'text' : news_data['data'], \
                             'category': news_data.target})
news_data_df.head()
```

The preceding code generates the following output:

	text	category
0	From: Steve@Busop.cit.wayne.edu (Steve Teolis)...	0
1	From: jks2x@holmes.acc.Virginia.EDU (Jason K. ...	0
2	From: wayne@uva386.schools.virginia.edu (Tony ...	1
3	From: lihan@ccwf.cc.utexas.edu (Bruce G. Bostw...	1
4	From: myoakam@cis.ohio-state.edu (micah r yoak...	0

Figure 3.50: Texts of news data as a DataFrame

6. Now, use the **lambda** function to extract tokens from each "text" of the **news_data_df** DataFrame. Check whether any of these tokens are stop words, lemmatize them, and concatenate them side by side. Use the **join** function to concatenate a list of words into a single sentence. Use the regular expression method (**re**) to replace anything other than letters, digits, and whitespaces with blank spaces. Add the following code to implement this:

```
news_data_df['cleaned_text'] = news_data_df['text']\
                        .apply(lambda x : ' '.join\
                        ([lemmatizer.lemmatize\
                          (word.lower()) \
                        for word in word_tokenize\
                        (re.sub(r'([^\s\w]|_)+', ' ', \
                         str(x))) if word.lower() \
                         not in stop_words]))
```

7. Add the following lines of code used to create a TFIDF matrix and transform it into a DataFrame:

```
tfidf_model = TfidfVectorizer(max_features=20)
tfidf_df = pd.DataFrame(tfidf_model.fit_transform\
            (news_data_df['cleaned_text']).todense())
tfidf_df.columns = sorted(tfidf_model.vocabulary_)
tfidf_df.head()
```

The preceding code generates the following output:

	00	article	com	edu	good	host	know	like	line	new	nntp	one	organization	posting	sale
0	0.719664	0.000000	0.000000	0.191683	0.0	0.124066	0.000000	0.153294	0.066931	0.520927	0.124370	0.0	0.068809	0.120711	0.161624
1	0.000000	0.000000	0.000000	0.219265	0.0	0.000000	0.353598	0.350704	0.153124	0.000000	0.000000	0.0	0.157421	0.000000	0.739523
2	0.000000	0.000000	0.000000	0.853563	0.0	0.000000	0.000000	0.000000	0.298044	0.000000	0.000000	0.0	0.306407	0.000000	0.000000
3	0.000000	0.267175	0.255208	0.567867	0.0	0.245034	0.000000	0.302760	0.132190	0.000000	0.245634	0.0	0.135900	0.238407	0.000000
4	0.000000	0.000000	0.000000	0.411807	0.0	0.266541	0.000000	0.000000	0.143793	0.000000	0.267194	0.0	0.147828	0.259333	0.694459

Figure 3.51: TFIDF representation as a DataFrame

8. Calculate the correlation matrix for this TFIDF representation. Add the following code to implement this:

```
correlation_matrix = tfidf_df.corr()
correlation_matrix.head()
```

The preceding code generates the following output:

	00	article	com	edu	good	host	know	like	line	new	nntp	one	organization
00	1.000000	-0.113080	-0.081874	-0.116847	-0.053495	-0.078405	-0.096597	-0.084413	-0.161674	0.026696	-0.084632	-0.076635	-0.208121
article	-0.113080	1.000000	0.125853	0.076146	-0.008246	-0.055519	0.025570	-0.000201	-0.158956	-0.121483	-0.046249	0.029978	-0.201204
com	-0.081874	0.125853	1.000000	-0.471456	-0.016128	-0.178742	-0.036333	-0.037284	-0.110011	-0.071355	-0.175256	-0.037293	-0.084630
edu	-0.116847	0.076146	-0.471456	1.000000	-0.098067	0.242610	-0.100041	-0.103703	-0.043210	-0.059893	0.247395	-0.119432	0.023394
good	-0.053495	-0.008246	-0.016128	-0.098067	1.000000	-0.098199	0.025899	0.045106	-0.186943	-0.046803	-0.098198	0.074548	-0.166908

Figure 3.52: Correlation matrix

9. Now, plot the correlation matrix using seaborn's heatmap function. Add the following code to implement this:

```
import seaborn as sns
fig, ax = plt.subplots(figsize=(20, 20))
sns.heatmap(correlation_matrix,annot=True, fmt='.1g', \
            vmin=-1, vmax=1, center= 0, cmap= 'coolwarm')
```

The preceding code generates the following output:

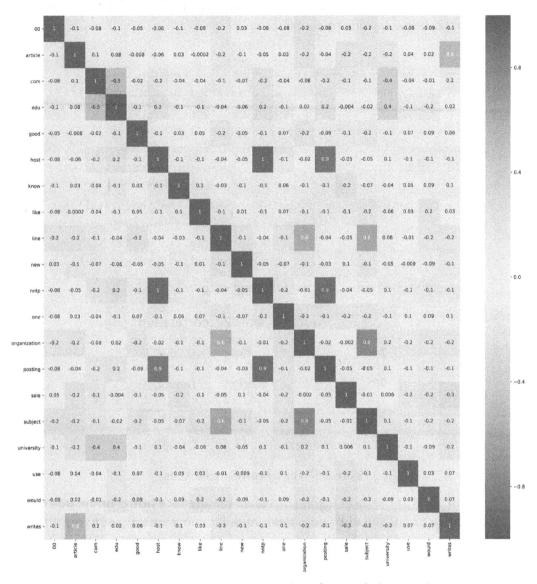

Figure 3.53: Heatmap representation of a correlation matrix

10. To identify a pair of terms with high correlation, create an upper triangular matrix from the correlation matrix. Create a stacked array out of it and traverse it. Add the following code to do this:

```
import numpy as np
correlation_matrix_ut = correlation_matrix.where(np.triu\
                        (np.ones(correlation_matrix.shape))\
                        .astype(np.bool))
correlation_matrix_melted = correlation_matrix_ut.stack()\
                        .reset_index()
correlation_matrix_melted.columns = ['word1', 'word2', \
                        'correlation']
correlation_matrix_melted[(correlation_matrix_melted['word1']\
                        !=correlation_matrix_melted['word2']) \
                        & (correlation_matrix_melted\
                        ['correlation']>.7)]
```

The preceding code generates the following output:

	word1	word2	correlation
95	host	nntp	0.953828
98	host	posting	0.896666
158	nntp	posting	0.934923
177	organization	subject	0.793946

Figure 3.54: Highly correlated tokens

You can see that the most highly correlated features are **host, nntp, posting, organization**, and **subject**. Next, we will remove **nntp, posting**, and **organization** since **host** and **subject** are highly correlated with them.

11. Remove terms for which the coefficient of correlation is greater than 0.7 and create a separate DataFrame with the remaining terms. Add the following code to do this:

```
tfidf_df_without_correlated_word = tfidf_df.drop(['nntp', \
                                    'posting', \
                                    'organization'],\
                                    axis = 1)

tfidf_df_without_correlated_word.head()
```

The preceding code generates the following output:

00	article	com	edu	good	host	know	like	line	new	one	sale
0.719664	0.000000	0.000000	0.191683	0.0	0.124066	0.000000	0.153294	0.066931	0.520927	0.0	0.1616
0.000000	0.000000	0.000000	0.219265	0.0	0.000000	0.353598	0.350704	0.153124	0.000000	0.0	0.7395
0.000000	0.000000	0.000000	0.853563	0.0	0.000000	0.000000	0.000000	0.298044	0.000000	0.0	0.0000
0.000000	0.267175	0.255208	0.567867	0.0	0.245034	0.000000	0.302760	0.132190	0.000000	0.0	0.0000
0.000000	0.000000	0.000000	0.411807	0.0	0.266541	0.000000	0.000000	0.143793	0.000000	0.0	0.6944

Figure 3.55: The DataFrame after removing correlated tokens

After removing the highly correlated words from the TFIDF DataFrame, it appears like this. We have cleaned the dataset to remove highly correlated features and are now one step closer to building our text classifier.

> **NOTE**
>
> To access the source code for this specific section, please refer to https://packt.live/39RdJTz.
>
> You can also run this example online at https://packt.live/2XbeCAX.

In the next section, we will learn how to reduce the size of the dataset and understand why this is necessary.

DIMENSIONALITY REDUCTION

There are some optional steps that we can follow on a case-to-case basis. For example, sometimes, the TFIDF matrix or Bag-of-Words representation of a text corpus is so big that it doesn't fit in memory. In this case, it would be necessary to reduce its dimension—that is, the number of columns in the feature matrix. The most popular method for dimension reduction is **Principal Component Analysis** (**PCA**).

PCA is used to perform dimensionality reduction. It converts a list of features (which may be correlated) into a list of variables that are linearly uncorrelated. These linearly uncorrelated variables are known as principal components. These principal components are arranged in descending order of the amount of variability they capture in the dataset. For example, let's consider a Twitter tweet dataset where people misspell words such as good and instead write "gud". PCA will combine these two highly correlated features into a single feature and reduce the dimensionality. In the next section, we'll look at an exercise to get a better understanding of this.

EXERCISE 3.12: PERFORMING DIMENSIONALITY REDUCTION USING PRINCIPAL COMPONENT ANALYSIS

In this exercise, we will reduce the dimensionality of a TFIDF matrix representation of sklearn's **fetch_20newsgroups** text dataset to two. Then, we'll create a scatter plot of these documents. Each category should be colored differently. Follow these steps to implement this exercise:

1. Open a Jupyter Notebook.

2. Insert a new cell and add the following code to import the necessary packages:

```
from sklearn.datasets import fetch_20newsgroups
import matplotlib as mpl
import pandas as pd
import numpy as np
import matplotlib.pyplot as plt
%matplotlib inline
import re
import string
from nltk import word_tokenize
from nltk.corpus import stopwords
from nltk.stem import WordNetLemmatizer
from sklearn.feature_extraction.text import TfidfVectorizer
from collections import Counter
from pylab import *
import nltk
nltk.download('stopwords')
nltk.download('punkt')
nltk.download('wordnet')
import warnings
warnings.filterwarnings('ignore')
```

3. Use stop words from the English language only. **WordNet** states the lemmatizer we will be using. Add the following code to implement this:

```
stop_words = stopwords.words('english')
stop_words = stop_words + list(string.printable)
lemmatizer = WordNetLemmatizer()
```

4. To specify the categories of news articles we want to fetch by, add the following code:

```
categories= ['misc.forsale', 'sci.electronics', \
             'talk.religion.misc']
```

5. To fetch sklearn's dataset corresponding to the categories we mentioned earlier, use the following lines of code:

```
news_data = fetch_20newsgroups(subset='train', \
                               categories=categories, \
                               shuffle=True, random_state=42, \
                               download_if_missing=True)
news_data_df = pd.DataFrame({'text' : news_data['data'], \
                             'category': news_data.target})
news_data_df.head()
```

The preceding code generates the following output:

	text	category
0	From: Steve@Busop.cit.wayne.edu (Steve Teolis)...	0
1	From: jks2x@holmes.acc.Virginia.EDU (Jason K. ...	0
2	From: wayne@uva386.schools.virginia.edu (Tony ...	1
3	From: llhan@ccwf.cc.utexas.edu (Bruce G. Bostw...	1
4	From: myoakam@cis.ohio-state.edu (micah r yoak...	0

Figure 3.56: News texts and their categories

6. Use the **lambda** function to extract tokens from each **text** item of the **news_data_df** DataFrame, check whether any of these tokens are stop words, lemmatize them, and concatenate them side by side. Use the **join** function to concatenate a list of words into a single sentence. Use the regular expression method (**re**) to replace anything other than letters, digits, and whitespaces with a blank space. Add the following code to implement this:

```
news_data_df['cleaned_text'] = news_data_df['text']\
                     .apply(lambda x : ' '.join\
                     ([lemmatizer.lemmatize\
                        (word.lower()) \
                     for word in word_tokenize\
                     (re.sub(r'([^\s\w]|_)+', ' ', \
                     str(x))) if word.lower() \
                     not in stop_words]))
```

7. The following lines of code are used to create a TFIDF matrix and transform it into a DataFrame:

```
tfidf_model = TfidfVectorizer(max_features=20)
tfidf_df = pd.DataFrame(tfidf_model.fit_transform\
        (news_data_df['cleaned_text']).todense())
tfidf_df.columns = sorted(tfidf_model.vocabulary_)
tfidf_df.head()
```

The preceding code generates the following output:

	00	article	com	edu	good	host	know	like	line	new	nntp	one
0	0.719664	0.000000	0.000000	0.191683	0.0	0.124066	0.000000	0.153294	0.066931	0.520927	0.124370	0.0
1	0.000000	0.000000	0.000000	0.219265	0.0	0.000000	0.353598	0.350704	0.153124	0.000000	0.000000	0.0
2	0.000000	0.000000	0.000000	0.853563	0.0	0.000000	0.000000	0.000000	0.298044	0.000000	0.000000	0.0
3	0.000000	0.267175	0.255208	0.567867	0.0	0.245034	0.000000	0.302760	0.132190	0.000000	0.245634	0.0
4	0.000000	0.000000	0.000000	0.411807	0.0	0.266541	0.000000	0.000000	0.143793	0.000000	0.267194	0.0

Figure 3.57: TFIDF representation as a DataFrame

8. In this step, we are using sklearn's **PCA** function to extract two principal components from the initial data. Add the following code to do this:

```
from sklearn.decomposition import PCA
pca = PCA(2)
pca.fit(tfidf_df)
reduced_tfidf = pca.transform(tfidf_df)
reduced_tfidf
```

The preceding code generates the following output:

```
array([[-0.18059713,  0.31553881],
       [-0.20040295,  0.37164609],
       [-0.4448022 , -0.0218631 ],
       ...,
       [-0.01981968,  0.32959235],
       [-0.37951601, -0.17487055],
       [-0.46927911, -0.02558393]])
```

Figure 3.58: Principal components

In the preceding screenshot, you can see the two principal components that the PCA algorithm calculated.

9. Now, we'll create a **scatter** plot using these principal components and represent each category with a separate color. Add the following code to implement this:

```
scatter = plt.scatter(reduced_tfidf[:, 0], \
                      reduced_tfidf[:, 1], \
                      c=news_data_df['category'], cmap='gray')
plt.xlabel('dimension_1')
plt.ylabel('dimension_2')
plt.legend(handles=scatter.legend_elements()[0], \
           labels=categories, loc='lower left')
plt.title('Representation of NEWS documents in 2D')
plt.show()
```

The preceding code generates the following output:

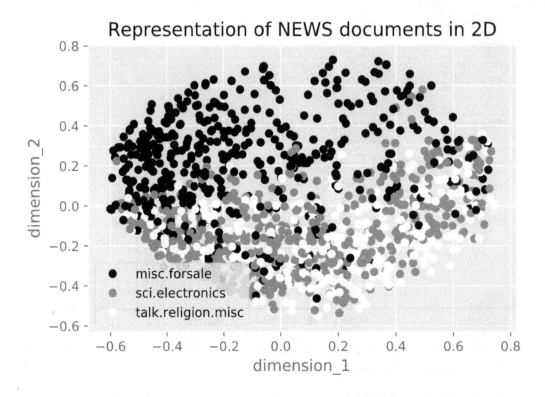

Figure 3.59: Two-dimensional representation of news documents

From the preceding plot, we can see that a scatter plot has been created in which each category of article is represented by a different color. This plot shows another important use case of dimensionality reduction: visualization. We were able to plot this two-dimensional image because we had two principal components. With the earlier TFIDF matrix, we had 20 features, which is impossible to visualize. In this section, you learned how to perform dimensionality reduction to save memory space and visualize datasets.

> **NOTE**
>
> To access the source code for this specific section, please refer to https://packt.live/2Xa5eh4.
>
> You can also run this example online at https://packt.live/3jU0dD7.

Next, we will learn how to evaluate the machine learning models that we train.

DECIDING ON A MODEL TYPE

Once the feature set is ready, it's necessary to decide on the type of model that will be used to deal with the problem. Usually, unsupervised models are chosen when data is not labeled. If we have a predefined number of clusters in mind, we go for clustering algorithms such as k-means; otherwise, we opt for hierarchical clustering. For labeled data, we generally follow supervised learning methods such as regression and classification.

If the outcome is continuous and numeric, we use regression. If it is discrete or categorical, we use classification. The Naive Bayes algorithm comes in handy for the fast development of simple classification models. More complex tree-based methods (such as decision trees, random forests, and so on) are needed when we want to achieve higher accuracy. In such cases, we sometimes compromise on model explainability and the time that's required to develop it. When the outcome of a model has to be the probability of the occurrence of a certain class, we use logistic regression.

EVALUATING THE PERFORMANCE OF A MODEL

Once a model is ready, it is necessary to evaluate its performance. This is because, without benchmarking it, we cannot be confident of how well or how badly it is functioning. It is not advisable to put a model into production without evaluating its efficiency. There are various ways to evaluate a model's performance. Let's work through them one by one:

- **Confusion Matrix**

  This is a two-dimensional matrix mainly used for evaluating the performance of classification models. Its columns consist of predicted values, and its rows consist of actual values. In other words, for a given confusion matrix, it is a crosstab between actual and predicted values. The cell entries denote how many of the predicted values match with the actual values, and how many don't. Consider the following image:

	Predicted 0	Predicted 1
Actual 0	TN	FP
Actual 1	FN	TP

  Figure 3.60: Confusion matrix

  In the preceding confusion matrix, the top-left cell will have the count of all correctly classified **0** values by the classifier, whereas the top-right cell will have the count of incorrectly classified **0** values, and so on. To create a confusion matrix using Python, you can use the following code:

```
from sklearn.metrics import confusion_matrix
confusion_matrix(actual_values,predicted_values)
```

- **Accuracy**

  Accuracy is defined as the ratio of correctly classified instances to the total number of instances. Whenever accuracy is used for model evaluation, we need to ensure that the data is balanced in terms of classes, meaning it should have an almost equal number of instances of each class. Let's use an example of a dataset that has 90% positive labels and 10% negative labels. A model that predicts all the data points as positive will receive 90% accuracy, but that will not be a good indicator of the performance of the model.

  To get the accuracy of the predicted values using Python, you can use the following code:

  ```
  from sklearn.metrics import accuracy_score
  accuracy_score(actual_values,predicted_values)
  ```

- **Precision and Recall**

  For a better understanding of precision and recall, let's consider a real-life example. If your mother tells you to explore the kitchen of your house, find items that need to be restocked, and bring them back from the market, you will bring P number of items from the market and show them to your mother. Out of P items, she finds Q items to be relevant. The Q/P ratio is called precision. However, in this scenario, she was expecting you to bring R items relevant to her. The ratio, Q/R, is referred to as recall:

  Precision = True Positive / (True Positive + False Positive)

  Recall = True Positive / (True Positive + False Negative)

- **F1 Score**

  For a given classification model, the F1 score is the harmonic mean of precision and recall. Harmonic mean is a way to find the average while giving equal weight to all numbers:

  *F1 score = 2 * ((Precision * Recall) / (Precision + Recall))*

  To get the F1 score, precision, and recall values using Python, you can use the following code:

  ```
  from sklearn.metrics import classification_report
  classification_report(actual_values,predicted_values)
  ```

- **Receiver Operating Characteristic (ROC) Curve**

  To understand the ROC curve, we need to get acquainted with the **True Positive Rate (TPR)** and the **False Positive Rate (FPR):**

  TPR = True Positive / (True Positive + False Negative)

  FPR = False Positive / (False Positive + True Negative)

  The output of a classification model can be probabilities. In that case, we need to set a threshold to obtain classes from those probabilities. The ROC curve is a plot between the TPR and FPR for various values of the threshold. The **area under the ROC** curve (**AUROC**) represents the efficiency of the model. The higher the AUROC, the better the model is. The maximum value of AUROC is 1. To create the ROC curve using Python, use the following code:

  ```
  fpr,tpr,threshold=roc_curve(actual_values, \
                          predicted_probabilities)
  print ('\nArea under ROC curve for validation set:', auc(fpr,tpr))
  fig, ax = plt.subplots(figsize=(6,6))
  ax.plot(fpr,tpr,label='Validation set AUC')
  plt.xlabel('False Positive Rate')
  plt.ylabel('True Positive Rate')
  ax.legend(loc='best')
  plt.show()
  ```

  Here, **actual_values** refers to the actual dependent variable values, whereas **predicted_probabilities** is the predicted probability of getting 1 from the trained predictor model.

- **Root Mean Square Error (RMSE)**

  This is mainly used for evaluating the accuracy of regression models. We define it as follows:

$$\mathrm{RMSE} = \sqrt{\frac{\sum(P_i - 0_i)^2}{n}}$$

Figure 3.61: Formula for root mean square error

Here, *n* is the number of samples, *Pi* is the predicted value of the *i*th observation, and *Oi* is the observed value of the *i*th observation. To find the RMSE using Python, use the following code:

```
from sklearn.metrics import mean_squared_error
rmse = sqrt(mean_squared_error(y_actual, y_predicted))
```

- **Maximum Absolute Percentage Error (MAPE)**

  Just like RMSE, this is another way to evaluate a regression model's performance. It is described as follows:

$$MAPE = \left(\frac{1}{n} \sum \frac{\left| O_i - P_i \right|}{\left| O_i \right|} \right) * 100 \text{ for all i from i to n}$$

Figure 3.62: Formula for maximum absolute percentage error

Here, *n* is the number of samples, *Pi* is the predicted value (that is, the forecast value) of the *ith* observation, and *Oi* is the observed value (that is, the actual value) of the *ith* observation. To find MAPE in Python, use the following code:

```
from sklearn.metrics import mean_absolute_error
mape = mean_absolute_error(y_actual, y_predicted) * 100
```

EXERCISE 3.13: CALCULATING THE RMSE AND MAPE OF A DATASET

In this exercise, we will calculate the RMSE and MAPE of hypothetical predicted and actual values. Follow these steps to implement this exercise:

1. Open a Jupyter Notebook.

2. Use sklearn's **mean_squared_error** to calculate the MSE and then use the **sqrt** function to calculate the RMSE. Add the following code to implement this:

```
from sklearn.metrics import mean_squared_error
from math import sqrt
y_actual = [0,1,2,1,0]
y_predicted = [0.03,1.2,1.6,.9,0.1]
rms = sqrt(mean_squared_error(y_actual, y_predicted))
print('Root Mean Squared Error (RMSE) is:', rms)
```

The preceding code generates the following output:

```
Root Mean Squared Error (RMSE) is: 0.21019038988498018
```

The preceding output shows the RMSE of the **y_actual** and **y_predicted** values that we created previously.

3. Next, use sklearn's **mean_absolute_error** to calculate the MAPE of a hypothetical model prediction. Add the following code to implement this:

```
from sklearn.metrics import mean_absolute_error
y_actual = [0,1,2,1,0]
y_predicted = [0.03,1.2,1.6,.9,0.1]
mape = mean_absolute_error(y_actual, y_predicted) * 100
print('Mean Absolute Percentage Error (MAPE) is:', \
        round(mape,2), '%')
```

The preceding code generates the following output:

```
Mean Absolute Percentage Error (MAPE) is 16.6 %
```

The preceding output shows the MAPE of the **y_actual** and **y_predicted** values that we created previously.

You have now learned how to evaluate the machine learning models that we train and are equipped to create your very own text classifier.

> **NOTE**
>
> To access the source code for this specific section, please refer to https://packt.live/3ggqRnp.
>
> You can also run this example online at https://packt.live/39E7i5S.

In the next section, we will solve an activity based on classifying text.

ACTIVITY 3.01: DEVELOPING END-TO-END TEXT CLASSIFIERS

For this activity, you will build an end-to-end classifier that figures out whether a news article is political or not.

> **NOTE**
>
> The dataset for this activity can be found at https://packt.live/39DUNHL.

Follow these steps to implement this activity:

1. Import the necessary packages.

2. Read the dataset and clean it.

3. Create a TFIDF matrix out of it.

4. Divide the data into training and validation sets.

5. Develop classifier models for the dataset.

6. Evaluate the models that were developed using parameters such as confusion matrix, accuracy, precision, recall, F1 plot curve, and ROC curve.

> **NOTE**
>
> The solution to this activity can be found on page 380.

We have seen how to build end-to-end classifiers. Developing an end-to-end classifier was done in phases. Firstly, the text corpus was cleaned and tokenized, features were extracted using TFIDF, and then the dataset was divided into training and validation sets. The XGBoost algorithm was used to develop a classification model. Finally, the performance was measured using parameters such as the confusion matrix, accuracy, precision, recall, F1 plot curve, and ROC curve. In the next section, you will learn how to build pipelines for NLP projects.

BUILDING PIPELINES FOR NLP PROJECTS

In general, a pipeline refers to a structure that allows a streamlined flow of air, water, or something similar. In this context, pipeline has a similar meaning. It helps to streamline various stages of an NLP project.

An NLP project is done in various stages, such as tokenization, stemming, feature extraction (TFIDF matrix generation), and model building. Instead of carrying out each stage separately, we create an ordered list of all these stages. This list is known as a pipeline. The **Pipeline** class of sklearn helps us combine these stages into one object that we can use to perform these stages one after the other in a sequence. We will solve a text classification problem using a pipeline in the next section to understand the working of a pipeline better.

EXERCISE 3.14: BUILDING THE PIPELINE FOR AN NLP PROJECT

In this exercise, we will develop a pipeline that will allow us to create a TFIDF matrix representation from sklearn's **fetch_20newsgroups** text dataset. Follow these steps to implement this exercise:

1. Open a Jupyter Notebook.

2. Insert a new cell and add the following code to import the necessary packages:

```
from sklearn.pipeline import Pipeline
from sklearn.feature_extraction.text import TfidfTransformer
from sklearn import tree
from sklearn.datasets import fetch_20newsgroups
from sklearn.feature_extraction.text import CountVectorizer
import pandas as pd
```

3. Specify the categories of news articles you want to fetch. Add the following code to do this:

```
categories = ['misc.forsale', 'sci.electronics', \
              'talk.religion.misc']
```

4. To fetch sklearn's **20newsgroups** dataset, corresponding to the categories mentioned earlier, we use the following lines of code:

```
news_data = fetch_20newsgroups(subset='train', \
                               categories=categories, \
                               shuffle=True, random_state=42, \
                               download_if_missing=True)
```

5. Define a pipeline consisting of two stages: **CountVectorizer** and **TfidfTransformer**. Fit it on the **news_data** we mentioned earlier and use it to transform that data. Add the following code to implement this:

```
text_classifier_pipeline = Pipeline([('vect', \
                                      CountVectorizer()), \
                                      ('tfidf', \
                                      TfidfTransformer())])
text_classifier_pipeline.fit(news_data.data, news_data.target)
pd.DataFrame(text_classifier_pipeline.fit_transform\
            (news_data.data, news_data.target).todense()).head()
```

The preceding code generates the following output:

	0	1	2	3	4	5	6	7	8	9	...	26016	26017	26018	26019	26020	26021	26022	26023
0	0.165523	0.000000	0.0	0.0	0.0	0.0	0.0	0.0	0.0	0.0	...	0.0	0.0	0.0	0.0	0.0	0.0	0.0	0.0
1	0.000000	0.000000	0.0	0.0	0.0	0.0	0.0	0.0	0.0	0.0	...	0.0	0.0	0.0	0.0	0.0	0.0	0.0	0.0
2	0.000000	0.000000	0.0	0.0	0.0	0.0	0.0	0.0	0.0	0.0	...	0.0	0.0	0.0	0.0	0.0	0.0	0.0	0.0
3	0.000000	0.000000	0.0	0.0	0.0	0.0	0.0	0.0	0.0	0.0	...	0.0	0.0	0.0	0.0	0.0	0.0	0.0	0.0
4	0.000000	0.081279	0.0	0.0	0.0	0.0	0.0	0.0	0.0	0.0	...	0.0	0.0	0.0	0.0	0.0	0.0	0.0	0.0

5 rows × 26026 columns

Figure 3.63: TFIDF representation of the DataFrame created using a pipeline

Here, we created a pipeline consisting of the count vectorizer and TFIDF transformer. The outcome of this pipeline is the TFIDF representation of the text data that has been passed to it as an argument.

> **NOTE**
>
> To access the source code for this specific section, please refer to https://packt.live/3gqpeUt.
>
> You can also run this example online at https://packt.live/3113qrJ.

SAVING AND LOADING MODELS

After a model has been built and its performance matches our expectations, we may want to save it for future use. This eliminates the time needed for rebuilding it. Models can be saved on the hard disk using the `joblib` and `pickle` libraries.

The `pickle` module makes use of binary protocols to save and load Python objects. `joblib` makes use of the `pickle` library protocols, but it improves on them to provide an efficient replacement to save large Python objects. Both libraries have two main functions that we will make use of to save and load our models:

- **dump**: This function is used to save a Python object to a file on the disk.

- **loads**: This function is used to load a saved Python object from a file on the disk.

To deploy saved models, we need to load them from the hard disk to the memory. In the next section, we will perform an exercise based on this to get a better understanding of this process.

EXERCISE 3.15: SAVING AND LOADING MODELS

In this exercise, we will create a TFIDF representation of sentences. Then, we will save this model on disk and later load it from the disk. Follow these steps to implement this exercise:

1. Open a Jupyter Notebook.

2. Insert a new cell and the following code to import the necessary packages:

```
import pickle
from joblib import dump, load
from sklearn.feature_extraction.text import TfidfVectorizer
```

3. Define a corpus consisting of four sentences by adding the following code:

```
corpus = ['Data Science is an overlap between Arts and Science',\
          'Generally, Arts graduates are right-brained and '\
          'Science graduates are left-brained', \
          'Excelling in both Arts and Science at a time '\
          'becomes difficult', \
          'Natural Language Processing is a part of Data Science']
```

4. Fit a TFIDF model to it. Add the following code to do this:

```
tfidf_model = TfidfVectorizer()
tfidf_vectors = tfidf_model.fit_transform(corpus).todense()
print(tfidf_vectors)
```

The preceding code generates the following output:

```
[[0.40332811 0.25743911 0.          0.25743911 0.          0.
  0.40332811 0.          0.          0.31798852 0.          0.
  0.          0.          0.          0.31798852 0.          0.
  0.          0.          0.40332811 0.          0.          0.
  0.42094668 0.          ]
 [0.          0.159139    0.49864399 0.159139    0.          0.
  0.          0.          0.49864399 0.          0.          0.
  0.24932199 0.49864399 0.          0.          0.          0.24932199
  0.          0.          0.          0.          0.          0.24932199
  0.13010656 0.          ]
 [0.          0.22444946 0.          0.22444946 0.35164346 0.35164346
  0.          0.35164346 0.          0.          0.35164346 0.35164346
  0.          0.          0.35164346 0.          0.          0.
  0.          0.          0.          0.          0.          0.
  0.18350214 0.35164346]
 [0.          0.          0.          0.          0.          0.
  0.          0.          0.          0.30887228 0.          0.
  0.          0.          0.          0.30887228 0.39176533 0.
  0.39176533 0.39176533 0.          0.39176533 0.39176533 0.
  0.2044394  0.          ]]
```

Figure 3.64: TFIDF representation as a matrix

5. Save this TFIDF model on disk using **joblib**. Add the following code to do this:

```
dump(tfidf_model, 'tfidf_model.joblib')
```

6. Finally, load this model from disk to memory and use it. Add the following code to do this:

```
tfidf_model_loaded = load('tfidf_model.joblib')
loaded_tfidf_vectors = tfidf_model_loaded.transform(corpus).todense()
print(loaded_tfidf_vectors)
```

The preceding code generates the following output:

```
[[0.40332811 0.25743911 0.          0.25743911 0.          0.
  0.40332811 0.          0.          0.31798852 0.          0.
  0.          0.          0.          0.31798852 0.          0.
  0.          0.          0.40332811 0.          0.          0.
  0.42094668 0.                     ]
 [0.          0.159139   0.49864399 0.159139   0.          0.
  0.          0.          0.49864399 0.          0.          0.
  0.24932199 0.49864399 0.          0.          0.          0.24932199
  0.          0.          0.          0.          0.          0.24932199
  0.13010656 0.                     ]
 [0.          0.22444946 0.          0.22444946 0.35164346 0.35164346
  0.          0.35164346 0.          0.          0.35164346 0.35164346
  0.          0.          0.35164346 0.          0.          0.
  0.          0.          0.          0.          0.          0.
  0.18350214 0.35164346]
 [0.          0.          0.          0.          0.          0.
  0.          0.          0.          0.30887228 0.          0.
  0.          0.          0.          0.30887228 0.39176533 0.
  0.39176533 0.39176533 0.          0.39176533 0.39176533 0.
  0.2044394  0.                     ]]
```

Figure 3.65: TFIDF representation as a matrix

7. Save this TFIDF model on disk using **pickle**. Add the following code to do this:

```
pickle.dump(tfidf_model, open("tfidf_model.pickle.dat", "wb"))
```

8. Load this model from disk to memory and use it. Add the following code to do this:

```
loaded_model = pickle.load(open("tfidf_model.pickle.dat", "rb"))
loaded_tfidf_vectors = loaded_model.transform(corpus).todense()
print(loaded_tfidf_vectors)
```

The preceding code generates the following output:

```
[[0.40332811 0.25743911 0.          0.25743911 0.          0.
  0.40332811 0.          0.          0.31798852 0.          0.
  0.          0.          0.          0.31798852 0.          0.
  0.          0.          0.40332811 0.          0.          0.
  0.42094668 0.          ]
 [0.          0.159139   0.49864399 0.159139   0.          0.
  0.          0.          0.49864399 0.          0.          0.
  0.24932199 0.49864399 0.          0.          0.          0.24932199
  0.          0.          0.          0.          0.          0.24932199
  0.13010656 0.          ]
 [0.          0.22444946 0.          0.22444946 0.35164346 0.35164346
  0.          0.35164346 0.          0.          0.35164346 0.35164346
  0.          0.          0.35164346 0.          0.          0.
  0.          0.          0.          0.          0.          0.
  0.18350214 0.35164346]
 [0.          0.          0.          0.          0.          0.
  0.          0.          0.          0.30887228 0.          0.
  0.          0.          0.          0.30887228 0.39176533 0.
  0.39176533 0.39176533 0.          0.39176533 0.39176533 0.
  0.2044394  0.          ]]
```

Figure 3.66: TFIDF representation as a matrix

From the preceding screenshot, we can see that the saved model and the model that was loaded from the disk are identical. You have now learned how to save and load models. This section marks the end of this chapter, where you learned how to build a text classifier from scratch.

> **NOTE**
>
> To access the source code for this specific section, please refer to https://packt.live/2BlDNmZ.
>
> You can also run this example online at https://packt.live/3hlay38.

SUMMARY

In this chapter, you learned about the different types of machine learning techniques, such as supervised and unsupervised learning. We explored unsupervised algorithms such as hierarchical clustering and k-means clustering, and supervised learning algorithms, such as k-nearest neighbor, the Naive Bayes classifier, and tree-based methods, such as random forest and XGBoost, that can perform both regression and classification. We discussed the need for sampling and went over different kinds of sampling techniques for splitting a given dataset into training and validation sets. Finally, we covered the process of saving a model on the hard disk and loading it back into memory for future use.

In the next chapter, you will learn about several techniques that you can use to collect data from various sources.

4

COLLECTING TEXT DATA WITH WEB SCRAPING AND APIS

OVERVIEW

This chapter introduces you to the concept of web scraping. You will first learn how to extract data (such as text, images, lists, and tables) from pages that are written using HTML. You will then learn about the various types of semi-structured data used to create web pages (such as JSON and XML) and extract data from them. Finally, you will use APIs for data extraction from Twitter, using the `tweepy` package.

INTRODUCTION

In the last chapter, we developed a simple classifier using feature extraction methods. We also covered different algorithms that fall under supervised and unsupervised learning. In this chapter, you will learn how to collect text data by scraping web pages, and then you will learn how to process that data. Web scraping helps you extract useful data from online content, such as product prices and customer reviews, which can then be used for market research, price comparison for products, or data analysis. You will also learn how to handle various kinds of semi-structured data, such as JSON and XML. We will cover different methods for extracting data using **Application Programming Interfaces** (**APIs**). Finally, we will explore different ways to extract data from different types of files.

COLLECTING DATA BY SCRAPING WEB PAGES

The basic building block of any web page is HTML (Hypertext Markup Language)—a markup language that specifies the structure of your content. HTML is written using a series of tags, combined with optional content. The content encompassed within HTML tags defines the appearance of the web page. It can be used to make words bold or italicize them, to add hyperlinks to the text, and even to add images. Additional information can be added to the element using attributes within tags. So, a web page can be considered to be a document written using HTML. Thus, we need to know the basics of HTML to scrape web pages effectively.

The following figure depicts the contents that are included within an HTML tag:

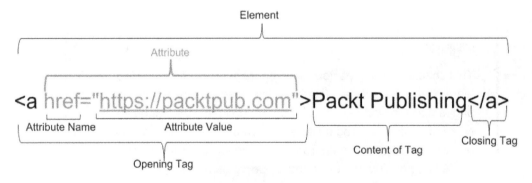

Figure 4.1: Tags and attributes of HTML

As you can see in the preceding figure, we can easily identify different elements within an HTML tag. The basic HTML structure and commonly used tags are shown and explained as follows:

```
1   <!DOCTYPE html>
2 ▾ <html>
3 ▾   <head>
4       <meta charset="utf-8">
5 ▾     <title>My Page</title>
6     </head>
7 ▾   <body>
8 ▾     <p>Hello, World!</p>
9     </body>
10  </html>
```

Figure 4.2: Basic HTML structure

- **DOCTYPE**: This is a must-have preamble for every HTML page. It informs the browser that the document is written in HTML.

- **<html>** tag: This is considered the root of the page, encompassing all of the page content. It is mainly divided into two tags—**<head>** and **<body>**.

- **<head>** tag: This tag provides meta-information about the web page.

- **<body>** tag: This tag comprises content such as text, image, tables, and lists.

- **<title>** tag: This sets the title of your page, which is what you'll see in the browser's tab.

- **<headline>** tag: As the name suggests, this represents six levels of section headings, from **<h1>** to **<h6>**.

- **<p>** tag: This is used to define the paragraph text content.

- **<i>** tag: We can use this tag to italicize the text.

- **<strong>** tag: This makes the text bold.

- **<li>** tag: We can use this tag to list the content in ordered (the **<ol>** tag) or unordered (the **<ul>** tag) list format.

- **<img>** tag: This tag is used to add an image in the HTML document.

- **<h1>** to **<h6>** tags: These represent the various levels of headings, with **<h1>** having the biggest size and **<h6>** having the smallest size.

- **<span>** tag: Although this tag provides no visual change by itself, it is useful for grouping inline-elements in a document and adding a hook to a part of a text or a part of a document.

- **<q>** tag: Quotes are written within the **<q>** tag in HTML.

- **table** tag: Tabular content is represented as a **table** tag, which contains **<th>** (table header), **<tr>** (table row), and **<td>** (table data).

- **<address>** tag: In HTML documents, addresses are enclosed within **<address>** tags.

In the next section, we will walk through an exercise in which we'll extract tag-based information from HTML files.

EXERCISE 4.01: EXTRACTION OF TAG-BASED INFORMATION FROM HTML FILES

In this exercise, we will extract addresses, quotes, text written in bold, and a table present in an HTML file.

> **NOTE**
>
> The data for this sample HTML file can be accessed from https://packt.live/338opvv.

Follow these steps to implement this exercise:

1. Open a Jupyter Notebook.

2. Insert a new cell and add the following code to import the **BeautifulSoup** library:

```
from bs4 import BeautifulSoup
```

BeautifulSoup is a Python library for pulling data out of HTML and XML files. It provides a parser for HTML/XML formats, allowing us to navigate, search, and modify the parsed tree.

3. Create an object of the **BeautifulSoup** class and pass the location of the HTML file to it:

```
soup = BeautifulSoup(open('../data/sample_doc.html'), \
                     'html.parser')
```

In the preceding line, **html.parser** is Python's built-in standard library parser. **BeautifulSoup** also supports third-party parsers such as **html5lib**, **lxml**, and others.

4. Add the following code to check the text contents of the **sample_doc.html** file:

```
soup.text
```

The preceding code generates the following output:

```
'\n\n\n A sample HTML Page \n\n\nI am staying at  Mess on No. 72, Banamali Naskar Lane, Kolkata. \nS
herlock  stays at 221B, Baker Street, London, UK. \nHamlet said to Horatio,   There are more things
in heaven and earth, Horatio,  Than are dreamt of in your philosophy. \n A table denoting details of
students\n\n\nname\nqualification\nadditional qualification\nother qualification\n\n\nGangaram\nB.Te
ch\nNA\nNA\n\n\nGanga\nB.A.\nNA\nNA\n\n\nRam\nB.Tech\nM.Tech\nNA\n\n\nRamlal\nB.Music\nNA\nDiploma i
n Music\n\n\n\n'
```

Figure 4.3: Text content of an HTML file

5. Similarly, to see the contents, you can simply write the following code:

```
soup.contents
```

```
['html', '\n', <html>
 <head>
 <title> A sample HTML Page </title>
 </head>
 <body>
 I am staying at <address> Mess on No. 72, Banamali Naskar Lane, Kolkata.</address> <br/>
 <b>Sherlock </b> stays at <address>221B, Baker Street, London, UK.</address> <br/>
 <b>Hamlet</b> said to <b>Horatio</b>, <br/> <q> There are more things in heaven and earth, Horatio,
 <br/> Than are dreamt of in your philosophy. </q>
```

Figure 4.4: Text content

6. To find the addresses from the document, insert a new cell and add the following code:

```
soup.find('address')
```

The preceding code generates the following output:

```
<address> Mess on No. 72, Banamali Naskar Lane, Kolkata.</address>
```

7. To locate all the **address** tags within the given content, write the following code:

```
soup.find_all('address')
```

The preceding code generates the following output:

```
[<address> Mess on No. 72, Banamali Naskar Lane, Kolkata.</address>,
 <address>221B, Baker Street, London, UK.</address>]
```

8. To find the quotes in the document, add the following code:

```
soup.find_all('q')
```

The preceding code generates the following output:

```
[<q> There are more things in heaven and earth, Horatio, <br/>
 Than are dreamt of in your philosophy. </q>]
```

9. To check all the bold items, write the following command:

```
soup.find_all('b')
```

The preceding code generates the following output:

```
[<b>Sherlock </b>, <b>Hamlet</b>, <b>Horatio</b>]
```

10. Write the following command to extract the tables in the document:

```
table = soup.find('table')
table
```

The preceding code generates the following output:

```
<table>
<tr>
<th>name</th>
<th>qualification</th>
<th>additional qualification</th>
<th>other qualification</th>
</tr>
<tr>
<td>Gangaram</td>
<td>B.Tech</td>
<td>NA</td>
<td>NA</td>
</tr>
<tr>
<td>Ganga</td>
<td>B.A.</td>
<td>NA</td>
<td>NA</td>
</tr>
```

Figure 4.5: Contents of the table tag

11. You can also view the contents of **table** by looping through it. Insert a new cell and add the following code to implement this:

```
for row in table.find_all('tr'):
    columns = row.find_all('td')
    print(columns)
```

The preceding code generates the following output:

```
[ ]
[<td>Gangaram</td>, <td>B.Tech</td>, <td>NA</td>, <td>NA</td>]
[<td>Ganga</td>, <td>B.A.</td>, <td>NA</td>, <td>NA</td>]
[<td>Ram</td>, <td>B.Tech</td>, <td>M.Tech</td>, <td>NA</td>]
[<td>Ramlal</td>, <td>B.Music</td>, <td>NA </td>, <td>Diploma in
Music</td>]
```

12. You can also locate specific content in the table. To locate the value at the intersection of the third row and the second column, write the following command:

```
table.find_all('tr')[3].find_all('td')[2]
```

The preceding code generates the following output:

```
<td>M.Tech</td>
```

We have learned how to extract tag-based information from an HTML file.

> **NOTE**
>
> To access the source code for this specific section, please refer to https://packt.live/3gekCAA.
>
> You can also run this example online at https://packt.live/2EyJp4q.

In the next section, we will focus on fetching content from web pages.

REQUESTING CONTENT FROM WEB PAGES

Whenever you visit a web page from your web browser, you are actually sending a request to fetch its content. This can be done using Python scripts. The Python **requests** package is widely used to handle all forms of HTTP requests. Let's walk through an exercise to get a better understanding of this concept.

To fetch content, you can use the **get()** method, which, as the name suggests, sends a **GET** request to the web page from which you want to fetch data. Let's perform a simple exercise now to get a better idea of how we can implement this in Python.

EXERCISE 4.02: COLLECTING ONLINE TEXT DATA

In this exercise, we will be fetching the web content with the help of **requests**. We will be pulling a text file from *Project Gutenberg*, the free e-book website, specifically, from the text file for Charles Dickens' famous book, *David Copperfield*. Follow these steps to complete this exercise:

1. Use the **requests** library to request the content of a book available online with the following set of commands:

```
import requests
"""
Let's read the text version of david copper field
available online
"""
r = requests.get('https://www.gutenberg.org/files/766/766-0.txt')
r.status_code
```

The preceding code generates the following output:

```
200
```

When the browser visits the website, it fetches the content of the specified URL. Similarly, using **requests**, we get the content from the specified URL and all the information gets stored in the **r** object. **200** indicates that we received the right response from the URL.

2. Locate the text content of the fetched file by using the **requests** object **r** and referring to the **text** attribute. Write the following code for this:

```
r.text[:1000]
```

The preceding code generates the following output:

```
'ï»¿The Project Gutenberg EBook of David Copperfield, by Charles Dickens\r\n\r\nThis eBook is for the use
of anyone anywhere at no cost and with\r\nalmost no restrictions whatsoever.  You may copy it, give it awa
y or\r\nre-use it under the terms of the Project Gutenberg License included\r\nwith this eBook or online a
t www.gutenberg.org\r\n\r\n\r\nTitle: David Copperfield\r\n\r\nAuthor: Charles Dickens\r\n\r\nRelease Dat
e: December, 1996  [Etext #766]\r\nPosting Date: November 24, 2009\r\nLast Updated: September 25, 2016\r\n
\r\nLanguage: English\r\n\r\nCharacter set encoding: UTF-8\r\n\r\n*** START OF THIS PROJECT GUTENBERG EBOO
K DAVID COPPERFIELD ***\r\n\r\n\r\n\r\n\r\n\r\nProduced by Jo Churcher\r\n\r\n\r\n\r\n\r\n\r\nDAVID COPPERFIEL
D\r\n\r\n\r\nBy Charles Dickens\r\n\r\n\r\n\r\n\r\n                    AFFECTIONATELY INSCRIBED TO\r\n
THE HON.  Mr. AND Mrs. RICHARD WATSON,\r\n                    OF ROCKINGHAM, NORTHAMPTONSHIRE.\r\n\r\n\r\n\r\nCONTE
NTS\r\n\r\n\r\n\r\n     I.      I Am Born\r\n      II.     I Observe\r\n      III.     I Have a Change\r\n        I
V.      I Fall into Disgrace\r\n      V.     '
```

Figure 4.6: Text contents of the file

3. Now, write the fetched content into a text file. To do this, add the following code:

```
from pathlib import Path
open(Path("../data/David_Copperfield.txt"),'w',\
     encoding='utf-8').write(r.text)
```

The preceding code generates the following output:

```
2033139
```

4. Similarly, we can do the same using Urllib3.First add the following code:

```
import urllib3
http = urllib3.PoolManager()
rr = http.request('GET', \
                  'http://www.gutenberg.org/files/766/766-0.txt')
rr.status
```

Again, we will get the output as **200**, similar to the previous method.

5. Add the following code to locate the text content:

```
rr.data[:1000]
```

You will see that you get the same output as shown in *Figure 4.6*.

6. Again, add the following code to write the fetched content into a text file:

```
open(Path("../data/David_Copperfield_new.txt"), \
     'wb').write(rr.data)
```

The preceding code will generate the following output:

```
2033139
```

We have just learned how to collect data from online sources with the help of the **requests** library.

> **NOTE**
>
> To access the source code for this specific section, please refer to https://packt.live/3fhu1pv.
>
> You can also run this example online at https://packt.live/2Dmov7L.

Now, let's look at analyzing HTML content from Jupyter Notebooks.

EXERCISE 4.03: ANALYZING THE CONTENT OF JUPYTER NOTEBOOKS (IN HTML FORMAT)

In this exercise, we will analyze the content of a Jupyter Notebook. We will count the number of images, list the packages that have been imported, and check the models and their performance.

> **NOTE**
>
> The HTML file used for this exercise, can be accessed at https://packt.live/3fcYIfJ.

Follow these steps to complete this exercise:

1. Import **BeautifulSoup** and pass the location of the given HTML file using the following commands:

```
from bs4 import BeautifulSoup

soup = BeautifulSoup(open('../data/text_classifier.html'), \
                     'html.parser')
soup.text[:100]
```

Here, we are loading HTML using **BeautifulSoup** and printing parsed content. The preceding code generates the following output:

```
'\n\n\nCh3_Activity7_Developing_end_to_end_Text_Classifiers\n\n\n\n
/*!\n*\n* Twitter Bootstrap\n*\n*/\n/*!\n*'
```

2. Use the **img** tag to count the number of images:

```
len(soup.find_all('img'))
```

The output shows that there are three **img** tags:

```
3
```

3. If you open the HTML file in the text editor or your web browser's console, you will see all **import** statements have the **class** attribute set to **nn**. So, to list all the packages that are imported, add the following code, referring to finding the **span** element with an **nn** class attribute:

```
[i.get_text() for i in soup.find_all\
('span',attrs={"class":"nn"})]
```

The preceding code generates the following output:

```
['pandas',
 'pd',
 'seaborn',
 'sns',
 'matplotlib.pyplot',
 'plt',
 're',
 'string',
 'nltk',
 'nltk.corpus',
 'nltk.stem',
 'sklearn.feature_extraction.text',
 'sklearn.model_selection',
 'pylab',
 'nltk',
 'warnings',
 'sklearn.metrics',
 'sklearn.linear_model',
 'sklearn.ensemble',
 'xgboost']
```

Figure 4.7: List of libraries imported

4. To extract the models and their performance, look at the HTML document and see which **class** attribute the models and their performance belong to. You will see the **h2** and **div** tags with the **class** attribute **output_subarea output_stream output_stdout output_text**. Add the following code to extract the models:

```
for md,i in zip(soup.find_all('h2'), \
soup.find_all('div',\
attrs={"class":"output_subarea output_stream "\
    "output_stdout output_text"})):
    print("Model: ",md.get_text())
    print(i.get_text())
    print("--------------------------------------------------------------
\n\n\n")
```

The preceding code generates the following output:

```
Model:  Logistic Regression¶

confusion matrix:
 [[28705   151]
 [ 1663  1396]]

accuracy:  0.943161522794924

classification report:
               precision    recall  f1-score   support

           0       0.95      0.99      0.97     28856
           1       0.90      0.46      0.61      3059

   micro avg       0.94      0.94      0.94     31915
   macro avg       0.92      0.73      0.79     31915
weighted avg       0.94      0.94      0.93     31915


Area under ROC curve for validation set: 0.911224422146723
```

Figure 4.8: Models and their performance

So, in the preceding output, we have extracted a classification report from the HTML file using **BeautifulSoup** by referring to the **<h2>** and **<div>** tags.

> **NOTE**
>
> To access the source code for this specific section, please refer to https://packt.live/2PaM1Yk.
>
> You can also run this example online at https://packt.live/315liSk.

So far, we have seen how to get content from the web using the **requests** package, and in this exercise, we saw how to parse and extract the desired information. Next time you come across an article and want to extract certain information from it, you will be able to put these skills to use, instead of manually going over all of the content.

ACTIVITY 4.01: EXTRACTING INFORMATION FROM AN ONLINE HTML PAGE

In this activity, we will extract data about Rabindranath Tagore from the Wikipedia page about him.

> **NOTE**
>
> Rabindranath Tagore was a poet and musician from South Asia whose art has had a profound influence on shaping the cultural landscape of the region. He was also the first Indian to win the Nobel Prize for Literature, in 1913.

After extracting the data, we will analyze information from the page. This should include the list of headings in the *Works* section, the list of his works, and the list of universities named after him. Follow these steps to implement this activity:

1. Open a Jupyter Notebook.

2. Import the requests and **BeautifulSoup** libraries.

3. Fetch the Wikipedia page from https://en.wikipedia.org/wiki/Rabindranath_Tagore using the **get** method of the **requests** library.

4. Convert the fetched content into HTML format using an HTML parser.

5. Print the list of headings in the *Works* section.

6. Print the list of original works written by Tagore in Bengali.

7. Print the list of universities named after Tagore.

> **NOTE**
>
> The solution to this activity can be found on page 386.

We are now well-versed in extracting generic data from HTML pages. Let's perform another activity now, where we'll be using regular expressions.

ACTIVITY 4.02: EXTRACTING AND ANALYZING DATA USING REGULAR EXPRESSIONS

To perform this activity, you will extract data from Packt's website. The data to be extracted includes frequently asked questions (FAQs) and their answers, phone numbers for customer care services, and the email addresses for customer care services. Follow these steps to complete this activity:

1. Import the necessary libraries and extract data from https://www.packtpub.com/support/faq using the `requests` library.

2. Fetch questions and answers from the data.

3. Create a DataFrame consisting of questions and answers.

4. Fetch email addresses with the help of regular expressions.

5. Fetch the phone numbers, with the help of regular expressions.

> **NOTE**
>
> The solution to this activity can be found on page 388.

In this activity, we were able to fetch data from online sources and analyze it in various ways. Now that we are well-versed in scraping web pages with the help of HTML, in the next section, we will discuss how to scrape web pages with semi-structured data.

DEALING WITH SEMI-STRUCTURED DATA

We learned about various types of data in *Chapter 2, Feature Extraction Methods*. Let's quickly recapitulate what semi-structured data refers to. A dataset is said to be semi-structured if it is not in a row-column format but, if required, can be converted into a structured format that has a definite number of rows and columns. Often, we come across data that is stored as key-value pairs or embedded between tags, as is the case with **JSON (JavaScript Object Notation)** and **XML (Extensible Markup Language)** files. These are the most popularly used instances of semi-structured data.

JSON

JSON files are used for storing and exchanging data. JSON is human-readable and easy to interpret. Just like text files and CSV files, JSON files are language-independent. This means that different programming languages, such as Python, Java, and so on, can work with JSON files effectively. In Python, a built-in data structure called a **dictionary** is capable of storing JSON objects as is. Generally, data in JSON objects is present in the form of key-value pairs. The datatype of values of JSON objects must be any of the following:

- A string

- A number

- Another JSON object

- An array

- A boolean

- Null

NoSQL databases (such as MongoDB) store data in the form of JSON objects. Most APIs return JSON objects. The following figure depicts what a JSON file looks like:

```
{
  "stones":[
    {
      "name": "Space Stone",
      "movies": ["Thor", "Captain America", "The
        Avengers"]

    },
    {
      "name": "Mind Stone",
      "movies": ["The Avengers", "The Winter Soldier",
        "Age of Ultron", "Civil War"]
    },
    {
      "name": "Reality Stone",
      "movies": ["The Dark World"]
    },
    {
      "name": "Power Stone",
      "movies": ["Guardians of the Galaxy"]
    },
    {
      "name": "Time Stone",
      "movies": ["Dr. Strange"]
    },
    {
      "name": "Soul Stone"
    }
  ]
}
```

Figure 4.9: A sample JSON file

Often, the response we get when requesting a URL is in the form of JSON objects. To deal with a JSON file effectively, we need to know how to parse it. The following exercise throws light on this.

EXERCISE 4.04: WORKING WITH JSON FILES

In this exercise, we will extract details such as the names of students, their qualifications, and additional qualifications from a JSON file.

> **NOTE**
>
> The sample JSON file can be accessed at https://packt.live/2P6Zwrl.

Follow these steps to complete this exercise:

1. Open a Jupyter Notebook.

2. Insert a new cell and import **json**. Pass the location of the file mentioned using the following commands:

```
import json
from pprint import pprint

data = json.load(open('../data/sample_json.json'))
pprint(data)
```

In the preceding code, we are importing Python's built-in **json** module and loading the local JSON file using the standard I/O operation of Python. This turns JSON into the Python **dict** object. The preceding code generates the following output:

```
{'students': [{'name': 'Gangaram', 'qualification': 'B.Tech'},
              {'name': 'Ganga', 'qualification': 'B.A.'},
              {'additional qualification': 'M.Tech',
               'name': 'Ram',
               'qualification': 'B.Tech'},
              {'name': 'Ramlal',
               'other qualification': 'Diploma in Music',
               'qualification': 'B.Music'}]}
```

Figure 4.10: Dictionary form of the fetched data

3. To extract the names of the students, add the following code:

```
[dt['name'] for dt in data['students']]
```

The preceding code generates the following output:

```
['Gangaram', 'Ganga', 'Ram', 'Ramlal']
```

4. To extract their respective qualifications, enter the following code:

```
[dt['qualification'] for dt in data['students']]
```

The preceding code generates the following output:

```
['B.Tech', 'B.A.', 'B.Tech', 'B.Music']
```

5. To extract their additional qualifications, enter the following code. Remember, not every student will have additional qualifications. Thus, we need to check this separately. Add the following code to implement this:

```
[dt['additional qualification'] if 'additional qualification' \
in dt.keys() else None for dt in data['students']]
```

The preceding code generates the following output:

```
[None, None, 'M.Tech', None]
```

As JSON objects are similar to the dictionary data structure of Python, they are widely used on the web to send and receive data across web applications.

> **NOTE**
>
> To access the source code for this specific section, please refer to https://packt.live/33aSGKi.
>
> You can also run this example online at https://packt.live/315MekS.

Now that we have learned how to load JSON data, let's extract data using another format, called **Extensible Markup Language** (**XML**), which is also used by web apps and Word documents to store information.

XML

Just like HTML, XML is another kind of markup language that stores data in between tags. It is human-readable and extensible; that is, we have the liberty to define our own tags. Attributes, elements, and tags in the case of XML are similar to those of HTML. An XML file may or may not have a declaration. But, if it has a declaration, then that must be the first line of the XML file.

This declaration statement has three parts: **Version**, **Encoding**, and **Standalone**. **Version** states which version of the XML standard is being used; **Encoding** states the type of character encoding being used in this file; **Standalone** tells the parser whether external information is needed for interpreting the content of the XML file. The following figure depicts what an XML file looks like:

```xml
<?xml version="1.0"?>
<data>
    <country name="Liechtenstein">
        <rank>1</rank>
        <year>2008</year>
        <gdppc>141100</gdppc>
        <neighbor name="Austria" direction="E"/>
        <neighbor name="Switzerland" direction="W"/>
    </country>
    <country name="Singapore">
        <rank>4</rank>
        <year>2011</year>
        <gdppc>59900</gdppc>
        <neighbor name="Malaysia" direction="N"/>
    </country>
    <country name="Panama">
        <rank>68</rank>
        <year>2011</year>
        <gdppc>13600</gdppc>
        <neighbor name="Costa Rica" direction="W"/>
        <neighbor name="Colombia" direction="E"/>
    </country>
</data>
```

Figure 4.11: A sample XML file

An XML file can be represented as a tree called an XML tree. This XML tree begins with the root element (the parent). This root element further branches into child elements. Each element of the XML file is a node in the XML tree. Those elements that don't have any children are leaf nodes. The following figure clearly differentiates between an original XML file and a tree representation of an XML file:

```
<data>
    <country name="Liechtenstein">
        <rank>1</rank>
        <year>2008</year>
        <gdppc>141100</gdppc>
        <neighbor name="Austria" direction="E"/>
        <neighbor name="Switzerland" direction="W"/>
    </country>
    <country name="Singapore">
        <rank>4</rank>
        <year>2011</year>
        <gdppc>59900</gdppc>
        <neighbor name="Malaysia" direction="N"/>
    </country>
    <country name="Panama">
        <rank>68</rank>
        <year>2011</year>
        <gdppc>13600</gdppc>
        <neighbor name="Costa Rica" direction="W"/>
        <neighbor name="Colombia" direction="E"/>
    </country>
</data>
```

```
data ..
  country ..
      @name: Liechtenstein
    rank 1
    year 2008
    gdppc 141100
    neighbor
        @name: Austria
        @direction: E
    neighbor
        @name: Switzerland
        @direction: W
  country ..
      @name: Singapore
    rank 4
```

Figure 4.12: Comparison of an XML structure

XML files are somewhat similar in structure to HTML, with the main difference being that, in XML, we have custom tags rather than the fixed tags vocabulary like HTML. As we learned how to parse HTML using **BeautifulSoup** before, let's learn how to parse XML files in the following exercise.

EXERCISE 4.05: WORKING WITH AN XML FILE

In this exercise, we will parse an XML file and print the details from it, such as the names of employees, the organizations they work for, and the total salaries of all employees.

> **NOTE**
>
> The sample XML data file can be accessed here: https://packt.live/3hPCaDI.

Follow these steps to complete this exercise:

1. Open a Jupyter Notebook.

2. Insert a new cell, import **xml.etree.ElementTree**, and pass the location of the XML file using the following code:

```
import xml.etree.ElementTree as ET
tree = ET.parse('../data/sample_xml_data.xml')
root = tree.getroot()
root
```

The preceding code generates the following output:

```
<Element 'records' at 0.112291710>
```

3. To check the tag of the fetched element, type the following code:

```
root.tag
```

The preceding code generates the following output:

```
'records'
```

4. Look for the **name** and **company** tags in the XML and print the data enclosed within them:

```
for record in root.findall('record')[:20]:
    print(record.find('name').text, "---", \
        record.find('company').text)
```

The preceding code generates the following output:

```
Peter Brewer --- Erat Ltd
Wallace Pace --- Sed Nunc Industries
Arthur Ray --- Amet Faucibus Corp.
Judah Vaughn --- Nunc Quis Arcu Inc.
Talon Combs --- Leo Elementum Ltd
Hall Bruce --- Proin Non Massa Consulting
Ronan Grant --- Scelerisque Sed Inc.
Dennis Whitaker --- Scelerisque Neque Foundation
Bradley Oconnor --- Aliquet Corporation
Forrest Alvarez --- Et Eros Institute
Ignatius Meyers --- Facilisis Lorem Limited
Bert Randolph --- Facilisis LLP
Victor Stevenson --- Lacinia Vitae Sodales Incorporated
Jamal Cummings --- Litora Ltd
Samson Estrada --- Lacinia Vitae Sodales Industries
Ira Spencer --- Duis Associates
Kevin Henson --- Sagittis Limited
Melvin Mccarthy --- Ipsum Suspendisse Company
Kieran Underwood --- Quisque Porttitor Eros Ltd
Cedric Phelps --- Lorem Vehicula Corp.
```

Figure 4.13: Data of the name and company tags printed

5. To find the sum of the salaries, create a list consisting of the salaries of all employees by iterating over each record and finding the **salary** tag in it. Next, remove the **$** and **,** from the string of salary content, and finally, type cast into the integer to get the sum at the end. Add the following code to do so:

```
sum([int(record.find('salary').text.replace('$','').\
replace(',','')) for record in root.findall('record')])
```

The preceding code generates the following output:

```
745609
```

Thus, we can see that the sum of all the salaries is $745,609. We just learned how to extract data from a local XML file. When we request data, many URLs return an XML file.

> **NOTE**
>
> To access the source code for this specific section, please refer to https://packt.live/3hQzuFM.
>
> You can also run this example online at https://packt.live/3jU8VRP.

In the next section, we will look at how APIs can be used to retrieve real-time data.

USING APIS TO RETRIEVE REAL-TIME DATA

API stands for **Application Programming Interface**. To understand what an API is, let's consider a real-life example. Suppose you have a socket plug in the wall, and you need to charge your cellphone using it. How will you do it? You will have to use a charger/adapter, which will enable you to connect the cellphone to the socket. Here, this adapter is acting as a mediator that connects the cellphone and the socket, thus enabling the smooth transfer of electricity between them.

Similarly, some websites do not provide their data directly. Instead, they provide APIs, which we can use to extract data from the websites. Just like the cellphone charger, an API acts as a mediator, enabling the smooth transfer of data between those websites and us. Let's perform a simple exercise to get hands-on experience of collecting data using APIs.

EXERCISE 4.06: COLLECTING DATA USING APIS

In this exercise, we will use the Currency Exchange Rates API to convert USD to another currency rate. Follow these steps to implement this exercise:

1. Open a Jupyter Notebook.

2. Import the necessary packages. Add the following code to do so:

```
import json
import pprint
import requests
import pandas as pd
```

3. Load the **json** data. Add the following code to do this:

```
r = requests.get("https://api.exchangerate-api.com/"\
                 "v4/latest/USD")
data = r.json()
pprint.pprint(data)
```

> **NOTE**
>
> Watch out for the slashes in the string below. Remember that the backslashes (\ ) are used to split the code across multiple lines, while the forward slashes (/) are part of the URL.

The preceding code generates the following output:

```
{'base': 'USD',
 'date': '2020-01-26',
 'rates': {'AED': 3.672058,
           'ARS': 60.073152,
           'AUD': 1.462619,
           'BGN': 1.772324,
           'BRL': 4.175311,
           'BSD': 1,
           'CAD': 1.313949,
           'CHF': 0.970542,
           'CLP': 775.032232,
           'CNY': 6.937035,
           'COP': 3356.26087,
           'CZK': 22.774105,
           'DKK': 6.769282,
           'DOP': 53.200551,
```

Figure 4.14: Fetched data in the Python dict format

4. To create the DataFrame of the fetched data and print it, add the following code:

```
df = pd.DataFrame(data)
df.head()
```

The preceding code generates the following output:

	base	date	time_last_updated	rates
AED	USD	2020-01-26	1579997437	3.672058
ARS	USD	2020-01-26	1579997437	60.073152
AUD	USD	2020-01-26	1579997437	1.462619
BGN	USD	2020-01-26	1579997437	1.772324
BRL	USD	2020-01-26	1579997437	4.175311

Figure 4.15: DataFrame showing details of currency exchange rates

Note that you will get a different output depending on the present currency exchange rates. We just learned how to collect data using APIs.

> **NOTE**
>
> To access the source code for this specific section, please refer to https://packt.live/3jQAcEG.
>
> You can also run this example online at https://packt.live/3jVIBa0.

In the next section, we will see how to create an API.

EXTRACTING DATA FROM TWITTER USING THE OAUTH API

Many popular websites, such as Twitter, provide an API that allows access to parts of their services so that people can build software that integrates with the website. We'll be focusing mainly on Twitter in this section. Twitter's data and services (such as tweets, advertisements, direct messages, and much more) can be accessed via the Twitter API. The Twitter API requires authentication and authorization to interact with its services using the **OAuth** method. Authentication is required to prove identity, while authorization proves the right to access its services and data. To access Twitter data and services using an API, you would need to register using a Twitter developer account.

You can collect data from Twitter using their Python module, named **Tweepy**. **Tweepy** is a Python library for accessing the Twitter API. It is great for simple automation and creating Twitter bots. It provides abstraction to communicate with Twitter and use its API to ease interactions, which makes this approach more efficient than using the **requests** library and Twitter API endpoints.

To use the **Tweepy** library, simply go to https://dev.twitter.com/apps/new and fill in the form; you'll need to complete the necessary fields, such as **App Name**, **Website URL**, **Callback URL**, and **App Usage**. Once you've done this, submit and receive the keys and tokens, which you can use for extracting tweets and more. However, before you do any of this, you'll first need to import the **tweepy** library.

Your Python code should look like this:

```
import tweepy

consumer_key = 'your consumer key here'
consumer_secret = 'your consumer secret key here'
access_token = 'your access token here'
access_token_secret = 'your access token secret here'

auth = tweepy.OAuthHandler(consumer_key, consumer_secret)
auth.set_access_token(access_token, access_token_secret)
api = tweepy.API(auth)
```

The preceding code uses **auth** instantiation from **OAuthHandler**, which takes in our consumer token and secret keys that were obtained during app registration. **OAuthHandler** handles interaction with Twitter's **OAuth** system.

To search for a query named **randomquery** using **tweepy**, you can use the **Cursor** object as follows:

```
tweepy.Cursor(api.search, q='randomquery', lang="en")
```

Cursor handles all the iterating-over-pages work for us behind the scenes, whereas the **api.search** method provides tweets that match a specified query given with the **q** parameter.

Let's do an activity now, to put our knowledge into practice.

ACTIVITY 4.03: EXTRACTING DATA FROM TWITTER

In this activity, you will extract 100 tweets containing the hashtag *#climatechange* from Twitter, using the Twitter API via the tweepy library, and load them into a pandas DataFrame. The following steps will help you implement this activity:

1. Log in to your Twitter account with your credentials.

2. Visit https://dev.twitter.com/apps/new and fill in the form by completing the necessary fields, such as `App Name`, providing `Website URL`, `Callback URL`, and `App Usage`.

3. Submit the form and receive the keys and tokens.

4. Use these keys and tokens in your application when making an API call for *#climatechange*.

5. Import the necessary libraries.

6. Fetch the data using the keys and tokens.

7. Create a DataFrame consisting of tweets.

> **NOTE**
>
> The full solution to this activity can be found on page 391.

In this activity, we extracted data from Twitter and loaded it into a pandas DataFrame. This data can also be used to analyze tweets and create a word cloud out of them, something that we will explore in detail in *Chapter 8*, *Sentiment Analysis*.

> **PUBLISHER'S NOTE**
>
> The preceding messages were extracted without bias from a given dataset and written by private individuals not affiliated with this company. The views expressed in these tweets do not necessarily reflect our company's official policies.

SUMMARY

In this chapter, we have learned various ways to collect data by scraping web pages. We also successfully scraped data from semi-structured formats such as JSON and XML and explored different methods of retrieving data in real time from a website without authentication. In the next chapter, you will learn about topic modeling—an unsupervised natural language processing technique that helps group documents according to the topics that it detects in them.

5

TOPIC MODELING

OVERVIEW

This chapter introduces topic modeling, which means using unsupervised machine learning to find "topics" within a given set of documents. You will explore the most common approaches to topic modeling, which are **Latent Semantic Analysis** (**LSA**), **Latent Dirichlet Allocation** (**LDA**), and the **Hierachical Dirichlet Process** (**HDP**), and learn the differences between them. You will then practice implementing these approaches in Python and review the common practical challenges in topic modeling. By the end of this chapter, you will be able to create topic models from any given dataset.

INTRODUCTION

In the previous chapter, we learned about different ways to collect data from local files and online resources. In this chapter, we will focus on **topic modeling**, which is an important area within natural language processing. Topic modeling is a simple way to capture the sense of what a document or a collection of documents is about. Note that in this case, documents are any coherent collection of words, which could be as short as a tweet or as long as an encyclopedia.

Topic modeling may be thought of as a way to automate the manual task of reading given document(s) to write an abstract, which you will then use to map the document(s) to a set of topics. Topic modeling is mostly done using unsupervised learning algorithms that detect topics on their own. Topic-modeling algorithms operate by performing statistical analysis on words or tokens in documents and using those statistics to automatically assign each document to multiple topics. A topic is represented by an arbitrary number and its keywords. When the topics are not interpretable, then topic modeling may be thought of as an automated process of a manual task in which the semantic structure or meaning of the documents was neither understood nor abstracted before mapping the document(s) to a set of topics.

Topic modeling generally uses unsupervised learning algorithms, as opposed to supervised learning algorithms. This means that, during training, we do not have to provide labels (that is, topic names corresponding to each document) in order to teach the model. This not only helps us discover interesting topics that might exist, but also reduces the manual effort spent in labeling texts. On the flip side, it can be a lot more challenging to evaluate the output of a topic model.

Topic modeling is often used as a first step to explore textual data in order to get a feel for the content of the text. This is especially true when abstracts/summaries are unavailable, and when the text is too large to be manually analyzed in the available timeframe.

TOPIC DISCOVERY

The main goal of topic modeling is to find a set of topics that can be used to classify a set of documents. These topics are implicit because we do not know what they are beforehand, and they are unnamed.

The number of topics could vary from around 3 to, say, 400 (or even more) topics. Since it is the algorithm that discovers the topics, the number is generally fixed as an input to the algorithm, except in the case of non-parametric models in which the number of topics is inferred from the text. These topics may not always directly correspond to topics that a human would find meaningful. In practice, the number of topics should be much smaller than the number of documents. In general, the number of topics specified in a parametric model ought to be greater than or equal to the expected number of topics in the text. In other words, one should err on the side of a greater number of topics rather than fewer topics. This is because fewer topics can cause a problem for the interpretability of topics. Also, the more documents that we provide, the better the algorithm can map the documents to non-mutually exclusive topics.

The number of topics chosen depends on the documents and the objectives of the project. You may want to increase the number of topics if you have a large number of documents or if the documents are fairly diverse. Conversely, if you are analyzing a narrow set of documents, you may want to decrease the number of topics. This generally flows from your assumptions about the documents. If you think that the document set might inherently contain a large number of topics, you should configure the algorithm to look for a similar number of topics.

EXPLORATORY DATA ANALYSIS

It is recommended to do exploratory data analysis prior to performing any machine learning project. This helps you learn about the probability distribution of the items in the dataset. We have seen this with word clouds in *Chapter 2, Feature Extraction Methods*. Even better exploration is possible with topic modeling. Doing this can give you a sense of the statistical properties of the text dataset and how the documents can be grouped.

For example, you might want to know whether the text dataset is skewed to any particular set of topics, or whether the sources are uniform or disparate. This data further allows us to choose the appropriate algorithms for the actual project.

TRANSFORMING UNSTRUCTURED DATA TO STRUCTURED DATA

Topic modeling clusters documents based on their topics. Specifically, it is a soft clustering method, as each document gets mapped to multiple topics. This is unlike hard clustering, which results in membership of an exemplar or a point of only one cluster. Topic models typically give a weight/probability of the document being associated with a topic.

Thus, you can have a matrix of documents by topic, wherein the intersection of a document and a topic refers to the weight/probability that the document is associated with the topic. This matrix is effectively a numeric representation of the text and can be considered a way to transform unstructured text into structured data. Such a transformation is also an example of dimensionality reduction, as unstructured text can have many more dimensions (each dimension corresponds to a unique word) than the number of dimensions in structured data (each dimension corresponds to a topic).

BAG OF WORDS

Before we explore modeling algorithms in depth, let's make a few simplifying assumptions. Firstly, we treat a document as a **bag of words**, meaning we ignore the structure and grammar of the document and just use the count of each word in the document to infer patterns in the variation of word counts. Ignoring the structure, sequences, and grammar allows us to use algorithms that rely on counts and probability to make the inferences.

As we have seen previously, a bag of words is a dictionary containing each unique word and the integer count of the occurrences of the word in the document. Like all models, it is, at best, an approximation of reality. All the topic-modeling algorithms that we will discuss consider the text as a bag of words.

> **NOTE**
>
> We will look at approaches that explicitly model sequences in later chapters. The sequential structure of languages is different from the sequential structure in time-series data. Moreover, some aspects of the sequential structure may be specific to the natural language being considered. This will be discussed in more detail in *Chapter 6, Vector Representation*.

TOPIC-MODELING ALGORITHMS

Topic-modeling algorithms operate on the following assumptions:

- Topics contain a set of words.

- Documents are made up of a set of topics.

Topics can be considered to be a weighted collection of words. After these common assumptions, different algorithms diverge in how they go about discovering topics. In the upcoming sections, we will cover in detail three topic-modeling algorithms—namely LSA, LDA, and HDP. Here, the term *latent* (the *L* in these acronyms) refers to the fact that the probability distribution of the topics is not directly observable. We can observe the documents and the words but not the topics.

> **NOTE**
>
> The LDA algorithm builds on the LSA algorithm. In this case, similar acronyms are indicative of this association.

LATENT SEMANTIC ANALYSIS (LSA)

We will start by looking at LSA. LSA actually predates the **World Wide Web**. It was first described in 1988. LSA is also known by an alternative name, **Latent Semantic Indexing** (**LSI**), particularly when it is used for semantic searches of document indices. The goal of LSA is to uncover the latent topics that underlie documents and words.

LSA — HOW IT WORKS

Consider that we have a collection of documents, and these documents are made up of words. Our goal is to discover the latent topics in the documents. So, in the beginning, we have a collection of documents that we can represent as a term-to-document matrix. This term-to-document matrix has terms as rows and documents as columns. The following table gives a simplified illustration of a term-to-document matrix:

Term to Document			
	Doc-1	Doc-1	Doc-3
Water			
Dog			
Willow			
Cart			
Pill			
Stone			

Figure 5.1: A simplified view of a term-to-document matrix

Now, we break this matrix down into three separate matrix factors, namely a term-to-topics matrix, a topic-importance matrix, and a topic-to-documents matrix. Let's consider the matrix shown on the left-hand side and the corresponding factor matrices on the right-hand side:

Figure 5.2: Document matrix and its broken matrices

As we can see in this diagram, the rectangular matrix is separated into the product of other matrices. The process takes a matrix, M, and splits it, as shown in the following formula:

$$M = U\Sigma V^{\top}$$

Figure 5.3: Splitting the matrix M

The following are the broad definitions of the preceding equation:

- M is an m×m matrix.

- U is an m×n matrix.

- Σ is an n×n diagonal matrix with non-negative real numbers.

- V is an m×n matrix.

- V^T is an n×m matrix, which is the transpose of V.

The matrices U and V^T are not unique as matrix factorization does not give unique factors. This is analogous to the fact that the number 108 can be factorized using three factors in more than one way: 9x1x12, 27x1x4, 3x1x36, and so on. In order to consistently get similar factors, a regularization parameter can be used. Moreover, the multiplication of the factor matrices gives a matrix approximately and not exactly equal to the original matrix. Collectively there are fewer elements in the factor matrices than in the original matrix and this is possible because the original matrix had many elements that were zero or close to zero.

The **gensim** library is a popular Python library for topic modeling. It is easy to use and provides various topic-modeling model classes, including **LdaModel** (for LDA) and **LsiModel** (for LSI).

The tomotopy library is also a powerful Python library for topic modeling. It too is easy to use and includes popular topic-modeling model classes, including **HDPModel** (for HDP) and **LDAModel** (for LDA).

Other Python topic-modeling libraries include scikit-learn and lda (for LDA).

KEY INPUT PARAMETERS FOR LSA TOPIC MODELING

We will be using the gensim library to perform LSA topic modeling. The key input parameters for gensim are `corpus`, the number of topics, and `id2word`. Here, the `corpus` is specified in the form of a list of documents in which each document is a list of tokens. The `id2word` parameter refers to a dictionary that is used to convert the corpus from a textual representation to a numeric representation such that each word corresponds to a unique number. Let's do an exercise to understand this concept better.

spaCy is a popular natural language processing Library for Python. In our exercises, we will be using spaCy to tokenize the text, lemmatize the tokens, and check which part-of-speech that token is. We will be using spaCy v2.1.3. After installing spaCy v2.1.3 we will need to download the English language model using the following code, so that we can load this model (since there are models for many different languages).

```
python -m spacy download en_core_web_sm
```

EXERCISE 5.01: ANALYZING WIKIPEDIA WORLD CUP ARTICLES WITH LATENT SEMANTIC ANALYSIS

In this exercise, you will perform topic modeling using LSA on a Wikipedia World Cup dataset. For this, you will make use of the `LsiModel` class provided by the gensim library. You will use the Wikipedia library to fetch articles, the spaCy engine for the tokenization of the text, and the newline character to separate documents within an article.

> **NOTE**
>
> The dataset used for this exercise can be found at https://packt.live/30dbExO.

Follow these steps to complete this exercise:

1. Open a Jupyter Notebook.

2. Insert a new cell and add the following code to import the necessary libraries:

```
import numpy as np
import matplotlib.pyplot as plt
%matplotlib inline
import pandas as pd
from gensim import corpora
from gensim.models import LsiModel
from gensim.parsing.preprocessing import preprocess_string
```

3. To clean the text, define a function to remove the non-alphanumeric characters and replace numbers with the **#** character. Replace instances of multiple newline characters with a single newline character. Use the newline character to separate out the documents in the corpus. Insert a new cell and add the following code to implement this:

```
import re

HANDLE = '@\w+'
LINK = 'https?://t\.co/\w+'
SPECIAL_CHARS = '&lt;|&lt;|&amp;|#'
PARA='\n+'
def clean(text):
    text = re.sub(LINK, ' ', text)
    text = re.sub(SPECIAL_CHARS, ' ', text)
    text = re.sub(PARA, '\n', text)
    return text
```

4. Insert a new cell and add the following code to find Wikipedia articles related to the World Cup:

```
import wikipedia
wikipedia.search('Cricket World Cup'),\
wikipedia.search('FIFA World Cup')
```

The code generates the following output:

```
(['Cricket World Cup',
  'Under-19 Cricket World Cup',
  '2019 Cricket World Cup',
  '2023 Cricket World Cup',
  '2015 Cricket World Cup',
  '2011 Cricket World Cup',
  '1996 Cricket World Cup',
  '2020 Under-19 Cricket World Cup',
  '1983 Cricket World Cup',
  "Women's Cricket World Cup"],
 ['2018 FIFA World Cup',
  'FIFA World Cup',
  '2022 FIFA World Cup',
  '2014 FIFA World Cup',
  '2010 FIFA World Cup',
  '2006 FIFA World Cup',
  '2026 FIFA World Cup',
  '2002 FIFA World Cup',
  '1998 FIFA World Cup',
  '1930 FIFA World Cup'])
```

Figure 5.4: Wikipedia articles related to the World Cup

5. Insert a new cell and add the following code fetch the Wikipedia articles about the 2018 FIFA World Cup and the 2019 Cricket World Cup, concatenate them, and show the result:

```
latest_soccer_cricket=['2018 FIFA World Cup',\
                       '2019 Cricket World Cup']
corpus=''
for cup in latest_soccer_cricket:
    corpus=corpus+wikipedia.page(cup).content
corpus
```

The code generates the following output:

```
'The 2018 FIFA World Cup was the 21st FIFA World Cup, an international football tournament contested by th
e men\'s national teams of the member associations of FIFA once every four years. It took place in Russia
from 14 June to 15 July 2018. It was the first World Cup to be held in Eastern Europe, and the 11th time t
hat it had been held in Europe. At an estimated cost of over $14.2 billion, it was the most expensive Worl
d Cup. It was also the first World Cup to use the video assistant referee (VAR) system.The finals involved
32 teams, of which 31 came through qualifying competitions, while the host nation qualified automatically.
Of the 32 teams, 20 had also appeared in the previous tournament in 2014, while both Iceland and Panama ma
de their first appearances at a FIFA World Cup. A total of 64 matches were played in 12 venues across 11 c
ities. Germany were the defending champions, but were eliminated in the group stage. Host nation Russia we
re eliminated in the quarter-finals.\nThe final took place on 15 July at the Luzhniki Stadium in Moscow, b
etween France and Croatia. France won the match 4-2 to claim their second World Cup title, marking the fou
rth consecutive title won by a European team.\n\n\n== Host selection ==\n\n\nThe bidding procedure to host t
```

Figure 5.5: Result after concatenating articles from 2018 and 2019

6. Insert a new cell and add the following code to clean the text, using the spaCy English language model to tokenize the corpus and exclude all tokens that are not detected as nouns:

```
text=clean(corpus)
import spacy
nlp = spacy.load('en_core_web_sm')
doc=nlp(text)
pos_list=['NOUN']
preproc_text=[]
preproc_sent=[]

for token in doc:
    if token.text!='\n':
        if not(token.is_stop) and not(token.is_punct) \
        and token.pos_ in pos_list:
            preproc_sent.append(token.lemma_)
    else:
        preproc_text.append(preproc_sent)
        preproc_sent=[]

#last sentence
preproc_text.append(preproc_sent)

print(preproc_text)
```

The code generates the following output:

```
[['football', 'tournament', 'man', 'team', 'member', 'association', 'year', 'place', 'time', 'cost', 'vide
o', 'assistant', 'referee', 'system', 'final', 'team', 'competition', 'host', 'nation', 'team', 'tournamen
t', 'appearance', 'total', 'match', 'venue', 'city', 'champion', 'group', 'stage', 'host', 'nation', 'quar
ter', 'final'], ['place', 'match', 'title', 'title', 'team'], ['host', 'selection'], ['bidding', 'procedur
e', 'tournament', 'association', 'interest', 'country', 'bid', 'proceeding', 'bid', 'government', 'lette
r', 'bid', 'bidding', 'process', 'nation', 'bid', 'nation', 'bid', 'bid', 'bid'], ['host', 'tournament',
```

Figure 5.6: Output after tokenizing the corpus

7. Insert a new cell and add the following code to convert the corpus into a list in which each token corresponds to a number for more efficient representation, as gensim requires it in this form. Then, find the topics in the corpus:

```
dictionary = corpora.Dictionary(preproc_text)
corpus = [dictionary.doc2bow(text) for text in preproc_text]
NUM_TOPICS=3
lsamodel=LsiModel(corpus, num_topics=NUM_TOPICS, \
                  id2word = dictionary)
lsamodel.print_topics()
```

The code generates the following output:

```
[(0,
  '0.554*"wicket" + 0.533*"run" + 0.288*"match" + 0.204*"tournament" + 0.177*"victory" + 0.169*"century" +
0.168*"over" + 0.138*"partnership" + 0.131*"score" + 0.127*"ball"'),
 (1,
  '0.444*"team" + 0.376*"match" + 0.356*"tournament" + 0.304*"time" + -0.246*"wicket" + -0.230*"run" + 0.1
71*"right" + 0.130*"country" + 0.124*"stage" + 0.124*"broadcast"'),
 (2,
  '-0.451*"match" + 0.389*"team" + -0.353*"right" + 0.315*"time" + -0.260*"broadcast" + -0.133*"viewer" +
-0.128*"rightsholder" + 0.117*"stage" + 0.115*"final" + 0.114*"nation"')]
```

Figure 5.7: Topics in the corpus

To create our **LsiModel**, we had to decide up front how many topics we wanted. This would not necessarily match the number of topics that are actually in the corpus.

Note that, in the output, you can see that negative weights are associated with some words in a few topics. Also, the sum of the weights does not add up to one. The weights are not to be interpreted as probabilities. This makes it difficult to even mechanically view the topic as a probability distribution over words. Additionally, it may be observed that topic **0** is essentially about cricket even though the corpus includes both soccer and cricket. Topic **1** seems to be related to a sports broadcast. Topic **2** does not seem to be interpretable.

8. To determine which topics have the highest weight for a document, insert a new cell and add the following code:

```
model_arr = np.argmax(lsamodel.get_topics(),axis=0)
y, x = np.histogram(model_arr, bins=np.arange(NUM_TOPICS+1))
fig, ax = plt.subplots()
plt.xticks(ticks=np.arange(NUM_TOPICS),\
           labels=np.arange(NUM_TOPICS+1))
ax.plot(x[:-1], y)
fig.show()
```

The code generates the following output:

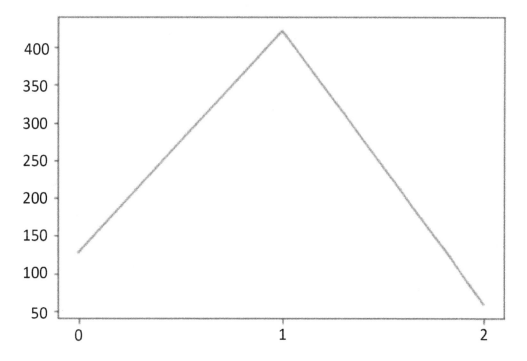

Figure 5.8: Graph representing weight of topics for the documents

We can see that topic **1** and topic **0** have the highest weight in almost all the documents.

> **NOTE**
>
> In general, the topics found are extremely sensitive to randomization in both gensim and tomotopy. While setting a `random_state` in gensim could help in reproducibility, in general, the topics found using tomotopy are superior from the perspective of interpretability. Generally, your output is expected to be different. In order to have exactly the same topic model, we can save and load topic models, which we'll do in *Exercise 5.04*, *Topics in The Life and Adventures of Robinson Crusoe by Daniel Defoe*.
>
> To access the source code for this specific section, please refer to https://packt.live/2PdOCkd.
>
> You can also run this example online at https://packt.live/3jSS7uB.

We have now performed topic modeling with the help of LSA. In the next section, we will learn about another topic-modeling algorithm: LDA. Before we move onto its implementation, let's quickly try and build a basic intuition about a couple of concepts that will help us in the subsequent sections.

DIRICHLET PROCESS AND DIRICHLET DISTRIBUTION

A Dirichlet process is a distribution over a distribution. It can be represented as $DP(\alpha, G)$ where G is the base distribution and α is the concentration parameter that defines how close $DP(\alpha, G)$ is to the base distribution G. It is for this reason that the Dirichlet process is a versatile way to represent various probability distributions. It is used for the HDP topic-modeling algorithm.

The Dirichlet distribution is a special case of the Dirichlet process, in which the number of topics needs to be specified explicitly. It is used for the LDA topic-modeling algorithm.

LATENT DIRICHLET ALLOCATION (LDA)

Instead of using matrix factorization, like we did for LSA, it is possible to consider a generative model called LDA. LDA is considered an advancement over probabilistic LSA. Probabilistic LSA is prone to overfitting as it does not probabilistically model the distribution of the documents. LDA is a three-level hierarchical generative statistical model that maps documents to topics, which in turn get mapped to words—all in a probabilistic way. In this case, we have two concentration parameters corresponding to the document level and the topic level.

LDA – HOW IT WORKS

To understand how LDA works, let's look at a simple example. We have four documents that contain only three unique words: **Cat**, **Dog**, and **Hippo**. The following figure shows the documents and the number of times each word is found in each document:

	Cat	Dog	Hippo
Document 1	10	0	0
Document 2	0	10	0
Document 3	0	0	10
Document 4	10	10	10

Figure 5.9: Occurrence of words in different documents

As we can see in the figure, the word **Cat** is found **10** times in **Document 1** and **Document 4** and **0** times in documents **2** and **3**. **Document 4** contains all three words **10** times each. For its analysis, LDA maintains two probability tables. The first table tracks the probability of selecting a specific word when sampling a specific topic. The second table keeps track of the probability of selecting a specific topic when sampling a particular document:

Words vs Topics

	Topic 1	Topic 2	Topic 3
Cat	0.00	0.00	0.99
Dog	0.99	0.00	0.00
Hippo	0.00	0.99	0.00

Documents vs Topics

	Topic 1	Topic 2	Topic 3
Document 1	0.030	0.030	0.939
Document 2	0.939	0.030	0.030
Document 3	0.030	0.939	0.030
Document 4	0.33	0.33	0.33

Figure 5.10: Probability tables

These probability tables reflect how likely it is to get a word if you sampled from each topic. If you sampled a word from **topic 3**, it would likely be **Cat** (probability 99%). If you sampled **Document 4**, then there is a one-third chance of getting each of the topics, since it contains all three t in equal proportions. In this example, a word is exclusive to a topic. In general, though, this is not the case.

The gensim and the scikit-learn libraries use one way of implementing LDA (called variational inference). The tomotopy and `lda` libraries use another way (called collapsed Gibbs sampling). It is essentially because of these differing implementations: when tomotopy is able to generate the topics in the available time, we usually prefer using tomotopy; otherwise we use gensim.

The parameters that we use for tomotopy are as follows:

- **corpus**: This refers to text that we want to analyze.

- Number of topics: This is the number of topics that the corpus contains.

- **iter**: This refers to the number of iterations that the model considers the corpus.

- α: This is associated with document generation.

- η: This is associated with topic generation.

- **seed**: This helps with fixing the initial randomization.

MEASURING THE PREDICTIVE POWER OF A GENERATIVE TOPIC MODEL

The predictive power of a generative topic model can be measured by analyzing the distribution of the generated corpus. Perplexity is a measure of how close the distribution of the words in the generated corpus is to reality. Log perplexity is a more convenient measure for this closeness. The formula for log perplexity is as follows:

$$
\log perplexity = -\frac{1}{n}\sum_{k=0}^{n}\log P(w)
$$

Figure 5.11: Formula for log perplexity

Here, *n* is number of words and *P(w)* is the probability associated with word *w*. We can see that negative log likelihood is identical to log perplexity.

Usually, a lower log perplexity means better performance. This is because the probability distribution of words is not uniform. It is concentrated on a small subset of words. And such a concentration (a non-uniform probability density function) causes a lower negative likelihood. In order to be sure that the model is generalizing well, the log likelihood should be computed on a hold-out sample. An extremely low negative log likelihood is indicative of an extremely low capacity of the model to learn. If a topic model has an unacceptable log perplexity on the corpus used for training then it is unlikely to perform well on a hold-out sample as it is indicative of the model having a low capacity to learn or it is indicative of the dataset not being generalizable. The negative log likelihood is approximately estimated in topic modeling libraries as it is intractable to calculate.

EXERCISE 5.02: FINDING TOPICS IN CANADIAN OPEN DATA INVENTORY USING THE LDA MODEL

In this exercise, we will use the tomotopy LDA model to analyze the Canadian Open Data Inventory. For simplicity, we will consider that the corpus has twenty topics.

> **NOTE**
>
> The dataset used for this exercise can be found at https://packt.live/2PbvMds.

The following steps will help you complete this exercise:

1. Open a Jupyter Notebook.

2. Insert a new cell and add the following code to import the necessary libraries:

```
import pandas as pd
pd.set_option('display.max_colwidth', 800)
import numpy as np
import matplotlib.pyplot as plt
%matplotlib inline
```

3. Insert a new cell and add the following code to read from a download of the Canadian Open Data Inventory, and clean the text:

```
OPEN_DATA_URL = '../data/canada-open-data/inventory.csv'

import re

HANDLE = '@\w+'
LINK = 'https?://t\.co/\w+'
SPECIAL_CHARS = '&lt;|&lt;|&amp;|#'
PARA='\n+'
def clean(text):
    text = re.sub(LINK, ' ', text)
    text = re.sub(SPECIAL_CHARS, ' ', text)
    text = re.sub(PARA, '\n', text)
    return text

catalog['description_en'].sample(frac=0.25,replace=False,\
                                 random_state=0).to_c \
                                 sv(OPEN_DATA_URL,\
                                 encoding='utf-8')
file='../data/canada-open-data/catalog.txt'
f=open(file,'r',encoding='utf-8')
text=f.read()
f.close()
text = clean(text)
```

4. Insert a new cell and add the following code to clean the text, using the spaCy English language model to tokenize the corpus and to exclude all tokens that are not detected as nouns:

```
import spacy
nlp = spacy.load('en_core_web_sm')
doc=nlp(text)
pos_list=['NOUN']
preproc_text=[]
preproc_sent=[]

for token in doc:
    if token.text!='\n':
        if not(token.is_stop) and not(token.is_punct) \
```

```
        and token.pos_ in pos_list:
            preproc_sent.append(token.lemma_)
    else:
        preproc_text.append(preproc_sent)
        preproc_sent=[]

#last sentence
preproc_text.append(preproc_sent)

print(preproc_text)
```

The code generates the following output:

```
[[], ['crop', 'residue', 'year', 'census'], ['investment', 'transit', 'infrastructure', 'city', 'communit
y', 'funding', 'environment', 'greenhouse', 'gas', 'emission', 'traffic', 'congestion', 'funding', 'provin
ce', 'territory', 'capita', 'basis'], ['need', 'background', 'soil', 'datum', 'assessment', 'site', 'regio
n', 'data', 'region', 'background', 'soil', 'concentration', 'metal', 'area', 'concentration', 'soil', 'qu
ality', 'guideline', 'jurisdiction', 'soil', 'database', 'database', 'region', 'background', 'soil', 'scre
ening', 'site', 'datum', 'background', 'range'], ['vegetable', 'storage', 'factory'], ['report', 'accoun
t'], ['facility', 'region', 'location', 'truck', 'trip', 'end', 'storage', 'handling', 'facility', 'busine
ss', 'dg'], ['park', 'pitcher', 'plant', 'morpology', 'availability', 'nitrogen', 'pitcher', 'plant', 'dev
elopment'], ['innovation', 'business', 'strategy', 'product', 'good', 'service', 'enterprise', 'market',
```

Figure 5.12: Tokenized corpus after text preprocessing

The pandas DataFrame was sampled. 25% of the dataset has been considered so that the memory restrictions related to spaCy can be addressed, since this is a fairly large sample.

5. Insert a new cell and add the following code to see how the negative log likelihood varies by the number of iterations:

```
import tomotopy as tp
NUM_TOPICS=20
mdl = tp.LDAModel(k=NUM_TOPICS,seed=1234)

for line in preproc_text:
    mdl.add_doc(line)

for i in range(0, 110, 10):
    mdl.train(i)
    print('Iteration: {}\tLog-likelihood: {}'.\
          format(i, mdl.ll_per_word))
```

The code generates the following output:

```
Iteration: 0    Log-likelihood: -11.093217577268552
Iteration: 10   Log-likelihood: -6.8822797912226115
Iteration: 20   Log-likelihood: -6.317129241581733
Iteration: 30   Log-likelihood: -6.157586638884254
Iteration: 40   Log-likelihood: -6.073628903605757
Iteration: 50   Log-likelihood: -6.0291570377492905
Iteration: 60   Log-likelihood: -6.005991344426762
Iteration: 70   Log-likelihood: -5.975599517879777
Iteration: 80   Log-likelihood: -5.959173736422274
Iteration: 90   Log-likelihood: -5.939598846671805
Iteration: 100  Log-likelihood: -5.935156891936913
```

Figure 5.13: Variation of negative log likelihood with different iterations

6. Insert a new cell and add the following code to train a topic model with ten iterations and to show the inferred topics:

```
mdl.train(10)
for k in range(mdl.k):
    print('Top 10 words of topic #{}'.format(k))
    print(mdl.get_topic_words(k, top_n=7))
```

The code generates the following output:

```
Top 10 words of topic #0
[('polygon', 0.36050185561180115), ('dataset', 0.0334757782722234726),
('information', 0.03004324994981289), ('soil', 0,029185116291046143),
('area', 0,026610717177391052), ('surface', 0.025752583518624306),
('map', 0.024036318063735962)]
```

7. Insert a new cell and add the following code to see the probability distribution of topics if you consider the entire dataset as a single document:

```
bag_of_words=[word for sent in preproc_text for word in sent]
doc_inst = mdl.make_doc(bag_of_words)
mdl.infer(doc_inst)[0]
np.argsort(np.array(mdl.infer(doc_inst)[0]))[::-1]
```

The code generates the following output:

```
array([11,17,14,19,12,  7,  4, 13, 10,  2,  3, 15,  1, 18, 16,  9,
0,
       6,  8,  5], dtype=int64)
```

8. Insert a new cell and add the following code to see the probability distribution of topic 11:

```
print(mdl.get_topic_words(11, top_n=7))
```

The code generates the following output

```
[('table', 0.24849626421928406), ('census', 0.1265643984079361),
('level', 0.06526772677898407), ('series', 0.06306280940771103),
('topic', 0.062401335686445236), ('geography', 0.062401335686445236),
('country', 0.06218084320425987)]
```

9. Insert a new cell and add the following code to see the probability distribution of topic 17:

```
print(mdl.get_topic_words(17, top_n=7))
```

The code generates the following output:

```
[('datum', 0.0603327676653862), ('information', 0.057247743010520935),
('year', 0.03462424501776695), ('dataset', 0.03291034325957298),
('project', 0.017828006289734993), ('website', 0.014057422056794167),
('activity', 0.012000739574432373)]
```

10. Insert a new cell and add the following code to see the probability distribution of topic 5:

```
print(mdl.get_topic_words(5, top_n=7))
```

The code generates the following output:

```
[('survey', 0.04966237023472786), ('catch', 0.03862873837351799),
('sponge', 0.0364220105111599), ('sea', 0.0342152863740921),
('datum', 0.028698472306132317), ('fishing', 0.02759511023759842),
('matter', 0.026491746306419373)]
```

Topic 11, topic 17, and topic 5 seem to be interpretable. One could say that topic 11, topic 17, and topic 5 seem to be broadly about geographical data, internet data, and marine life data respectively.

> **NOTE**
>
> In general, the topics found are extremely sensitive to randomization in both gensim and tomotopy. While setting a `random_state` in gensim could help in reproducibility, in general, the topics found using tomotopy are superior from the perspective of interpretability. Generally, your output is expected to be different. In order to have exactly the same topic model, we can save and load topic models; we do this in *Exercise 5.04*, *Topics in The Life and Adventures of Robinson Crusoe by Daniel Defoe*.
>
> To access the source code for this specific section, please refer to https://packt.live/33d0FGw.
>
> This section does not currently have an online interactive example and will need to be run locally.

ACTIVITY 5.01: TOPIC-MODELING JEOPARDY QUESTIONS

Jeopardy is a popular TV show that covers a variety of topics. In this show, participants are given answers and then asked to frame questions. The purpose of this activity is to give a real-world feel to some of the complexity associated with topic modeling. In this activity, you will do topic modeling on a dataset of Jeopardy questions.

> **NOTE**
>
> The dataset to be used for this activity can be found at https://packt.live/2PbvMds.

Follow these steps to complete this activity:

1. Open a Jupyter Notebook.

2. Insert a new cell and import pandas and other necessary libraries.

3. Load the dataset into a pandas DataFrame.

4. Clean the data by dropping the DataFrame rows where the **Question** column has empty cells.

5. Find the unique number of categories based on the **Category** column.

6. Randomly select 4% of the questions. Tokenize the text using spaCy. Select tokens that are nouns/verbs/adjectives or a combination.

7. Train a tomotopy LDA model with 1,000 topics.

8. Print the log perplexity.

9. Find the probability distribution on the entire dataset.

10. Sample a few topics and check for interpretability.

> **NOTE**
>
> The full solution to this activity can be found on page 395.

HIERARCHICAL DIRICHLET PROCESS (HDP)

HDP is a non-parametric variant of LDA. It is called "non-parametric" since the number of topics is inferred from the data, and this parameter isn't provided by us. This means that this parameter is learned and can increase (that is, it is theoretically unbounded).

The tomotopy HDP implementation can infer between 1 and 32,767 topics. gensim's HDP implementation seems to fix the number of topics at 150 topics. For our purposes, we will be using the tomotopy HDP implementation.

The gensim and the scikit-learn libraries use variational inference, while the tomotopy library uses collapsed Gibbs sampling. When the time required by collapsed Gibbs sampling is not an issue, then it is preferable to use collapsed Gibbs sampling over variational inference. In other cases, we may prefer to use variational inference. For the tomotopy library, the following parameters are used:

iter: This refers to the number of iterations that the model considers the corpus.

α: This concentration parameter is associated with document generation.

η: This concentration parameter is associated with topic generation.

seed: This fixes the initial randomization.

min_cf: This helps eliminate those words that occur fewer times than the frequency specified by us.

To get a better understanding of this, let's perform some simple exercises.

EXERCISE 5.03: TOPICS IN AROUND THE WORLD IN EIGHTY DAYS

In this exercise, we will make use of the tomotopy HDP model to analyze the text file for Jules Verne's *Around the World in Eighty Days*, available from the Gutenberg Project. We will use the **min_cf** hyperparameter that is used to ignore words that occur fewer times than the specified frequency and discuss its impact on the interpretability of topics.

> **NOTE**
>
> The dataset used for this exercise can be found at https://packt.live/2Xdv4kt.

1. Open a Jupyter Notebook.

2. Insert a new cell and add the following code to import the necessary libraries:

```
import pandas as pd
pd.set_option('display.max_colwidth', 800)
import numpy as np
import matplotlib.pyplot as plt
%matplotlib inline
```

3. Insert a new cell and add the following code to read from a download of the Gutenberg Project's *Around the World in Eighty Days* by Jules Verne, and clean the text:

```
OPEN_DATA_URL = '../data/aroundtheworld/pg103.txt'
f=open(OPEN_DATA_URL,'r',encoding='utf-8')
text=f.read()
f.close()

import re
```

```
HANDLE = '@\w+'
LINK = 'https?://t\.co/\w+'
SPECIAL_CHARS = '&lt;|&lt;|&amp;|#'
PARA='\n+'
def clean(text):
    text = re.sub(LINK, ' ', text)
    text = re.sub(SPECIAL_CHARS, ' ', text)
    text = re.sub(PARA, '\n', text)
    return text


text = clean(text)
text
```

The code generates the following output:

```
'\ufeffThe Project Gutenberg EBook of Around the World in 80 Days, by Jules Verne\nThis eBook is for the u
se of anyone anywhere at no cost and with\nalmost no restrictions whatsoever.  You may copy it, give it aw
ay or\nre-use it under the terms of the Project Gutenberg License included\nwith this eBook or online at w
ww.gutenberg.net\nTitle: Around the World in 80 Days\nAuthor: Jules Verne\nRelease Date: May 15, 2008 [EBo
ok  103]\nLast updated: February 18, 2012\nLast updated: May 5, 2012\nLanguage: English\n*** START OF THIS
PROJECT GUTENBERG EBOOK AROUND THE WORLD IN 80 DAYS ***\nAROUND THE WORLD IN EIGHTY DAYS\nCONTENTS\nCHAPTE
R\n    I  IN WHICH PHILEAS FOGG AND PASSEPARTOUT ACCEPT EACH OTHER, THE\n       ONE AS MASTER, THE OTH
ER AS MAN\n    II  IN WHICH PASSEPARTOUT IS CONVINCED THAT HE HAS AT LAST FOUND\n       HIS IDEAL\n
III  IN WHICH A CONVERSATION TAKES PLACE WHICH SEEMS LIKELY TO COST\n      PHILEAS FOGG DEAR\n    IV
IN WHICH PHILEAS FOGG ASTOUNDS PASSEPARTOUT, HIS SERVANT\n    V  IN WHICH A NEW SPECIES OF FUNDS, UNKNOW
```

Figure 5.14: Text from "Around the World in Eighty Days"

4. Insert a new cell and add the following code to import the necessary libraries, clean the text (using the spaCy English language model to tokenize the corpus), and exclude all tokens that are not detected as nouns:

```
import spacy
nlp = spacy.load('en_core_web_sm')
doc=nlp(text)
pos_list=['NOUN']

preproc_text=[]
preproc_sent=[]

for token in doc:
    if token.text!='\n':
        if not(token.is_stop) and not(token.is_punct) \
        and token.pos_ in pos_list:
            preproc_sent.append(token.lemma_)
    else:
        preproc_text.append(preproc_sent)
```

```
        preproc_sent=[]

preproc_text.append(preproc_sent) #last sentence

print(preproc_text)
```

The code generates the following output:

```
[[['ebook', 'day'], ['use', 'cost'], ['restriction'], ['term'], [], ['title', 'world', 'day'], ['author'],
['date'], [], [], [], ['START', 'ebook', 'world', 'day'], ['days'], [], ['chapter', 'fogg', 'PASSEPARTOU
T', 'PASSEPARTOUT', 'convinced', 'ideal', 'iii', 'conversation', 'takes', 'LIKELY', 'astound', 'PASSEPARTO
UT', 'v', 'species', 'fund', 'man', 'change', 'DETECTIVE', 'viii', 'PASSEPARTOUT', 'ix', 'prove', 'design
s', 'X', 'PASSEPARTOUT', 'xi', 'means', 'price', 'xii', 'companion', 'forest', 'ensued', 'PASSEPARTOUT',
'proof', 'fortune', 'favors', 'xiv', 'fogg', 'length', 'thinking', 'xv', 'bag', 'banknote', 'thousand', 'p
ound', 'xvi', 'fix', 'happened', 'voyage', 'xviii', 'FIX', 'xix', 'takes', 'interest', 'master', 'xx', 'xx
i', 'master', 'risk', 'losing', 'reward', 'pound', 'XXII', 'PASSEPARTOUT', 'money', 'XXIII', 'passepartou
t', 'xxiv', 'glimpse', 'xxvii', 'passepartout', 'undergoe', 'speed', 'mile', 'hour', 'course', 'history',
```

Figure 5.15: Tokenized corpus after the text is cleaned

5. Insert a new cell and add the following code to create HDP models in which tokens that occur fewer than five times are ignored, and then show how the negative log likelihood varies according to the number of iterations:

```
import tomotopy as tp
mdl = tp.HDPModel(min_cf=5,seed=0)

for line in preproc_text:
    mdl.add_doc(line)

for i in range(0, 100, 10):
    mdl.train(i)
    print('Iteration: {}\tLog-likelihood: {}'.\
          format(i, mdl.ll_per_word))

for k in range(mdl.k):
    print('Top 10 words of topic #{}'.format(k))
    print(mdl.get_topic_words(k, top_n=7))
```

The code generates the following output:

```
Iteration: 0    Log-likelihood: -6.929488035114073
Iteration: 10   Log-likelihood: -6.5324189802807116
Iteration: 20   Log-likelihood: -6.564015251147695
Iteration: 30   Log-likelihood: -6.90579612594132
Iteration: 40   Log-likelihood: -7.206281552678545
Iteration: 50   Log-likelihood: -7.352613202015137
Iteration: 60   Log-likelihood: -7.337697058844141
Iteration: 70   Log-likelihood: -7.364130322712162
Iteration: 80   Log-likelihood: -7.329386253940604
Iteration: 90   Log-likelihood: -7.356350772885065
```

Figure 5.16: Variation of negative log likelihood with number of iterations

6. Insert a new cell and add the following code to see the probability distribution of topics if you consider the entire dataset as a single document:

```
bag_of_words=[word for sent in preproc_text for word in sent]
doc_inst = mdl.make_doc(bag_of_words)
np.argsort(np.array(mdl.infer(doc_inst)[0]))[::-1]
```

The code generates the following output:

```
array([ 33,  21,  70,  82,  68,  80,  69,  81,  83,  71,  72,  84,  23,
        35,  22,  34,  32,  20, 106, 165,  60,  48,  12, 228, 164, 138,
        96, 116, 276, 324, 310, 278, 198, 166, 244,  62,  14,  98, 352,
       196, 292,   0,   1, 180,  41,  40,  89,  88, 354, 184, 296, 104,
       312, 136, 248, 216, 168, 120, 232, 264, 280, 200, 152, 328,  17,
        65,  95, 146, 194, 258, 306, 114, 322, 226, 130, 178, 242, 274,
       290, 162, 338, 210, 308, 148, 132, 250, 377, 364, 345, 348, 374,
```

Figure 5.17: Probability distribution of topics if the entire dataset is considered

7. Insert a new cell and add the following code to see the probability distribution of topic 33:

```
print(mdl.get_topic_words(33, top_n=7))
```

The code generates the following output:

```
[('danger', 0.15349544458475113), ('hour', 0.0015197568573057652),
('time', 0.0015197568573057652), ('train', 0.0015197568573057652),
('master', 0.0015197568573057652), ('man', 0.0015197568573057652),
('steamer', 0.0015197568573057652)]
```

8. Insert a new cell and add the following code to see the probability distribution of topic 21:

```
print(mdl.get_topic_words(21, top_n=7))
```

The code generates the following output:

```
[('hour', 0.1344495415687561), ('minute', 0.1232500821352005),
('day', 0.08405196666717529), ('quarter', 0.07285250723361969),
('moment', 0.07285250723361969), ('clock', 0.005605331063270569),
('card', 0.039254117757081985)]
```

9. Insert a new cell and add the following code to see the probability distribution of topic 70:

```
print(mdl.get_topic_words(70, top_n=7))
```

The code generates the following output:

```
[('event', 0.12901155650615692), ('midnight', 0.12901155650615692),
('detective', 0.06482669711112976), ('bed', 0.06482669711112976),
('traveller', 0.06482669711112976), ('watch', 0.06482669711112976),
('clown', 0.06482669711112976)]
```

10. Insert a new cell and add the following code to see the probability distribution of topic 4:

```
print(mdl.get_topic_words(4, top_n=7))
```

The code generates the following output:

```
[('house', 0.20237493515014648), ('opium', 0.10131379961967468),
('town', 0.07604850828647614), ('brick', 0.07604850828647614),
('mansion', 0.07604850828647614), ('glimpse', 0.50783220678567886),
('ball', 0.050783220678567886)]
```

We can see that ignoring tokens that occur fewer than five times significantly improves the interpretability of the topic model. Also, we have 378 topics in all, many of which are not likely to be interpretable. So, what does this mean? Let's analyze a corpus from another classic and then return to these questions.

> **NOTE**
>
> In general, the topics found are extremely sensitive to randomization in both gensim and tomotopy. While setting a `random_state` in gensim could help reproducibility, the topics found using tomotopy are superior from the perspective of interpretability. Your output is expected to be different. In order to have exactly the same topic model, we can save and load topic models, which we'll do now in *Exercise 5.04*, *Topics in The Life and Adventures of Robinson Crusoe by Daniel Defoe*.
>
> To access the source code for this specific section, please refer to https://packt.live/3jTxUVk.
>
> You can also run this example online at https://packt.live/2X8lG1p.

EXERCISE 5.04: TOPICS IN THE LIFE AND ADVENTURES OF ROBINSON CRUSOE BY DANIEL DEFOE

In this exercise, we will make use of the tomotopy HDP model to analyze a text corpus taken from the text file for Daniel Defoe's *The Life and Adventures of Robinson Crusoe*, available on the Gutenberg Project website. Here, we will take the value of α as 0.8 and experiment with selecting tokens based on different combinations of parts of speech, before training the model.

> **NOTE**
>
> The dataset used for this exercise can be found at https://packt.live/3ffhfrP.

1. Open a Jupyter Notebook.

2. Insert a new cell and add the following code to import the necessary libraries:

```
import pandas as pd
pd.set_option('display.max_colwidth', 800)
import numpy as np
import matplotlib.pyplot as plt
%matplotlib inline
```

3. Insert a new cell and add the following code to read from a download of the Gutenberg Project's *The Life and Adventures of Robinson Crusoe* by Daniel Defoe, and clean the text:

```
OPEN_DATA_URL = '../data/robinsoncrusoe/521-0.txt'
f=open(OPEN_DATA_URL,'r',encoding='utf-8')
text=f.read()
f.close()

import re

HANDLE = '@\w+'
LINK = 'https?://t\.co/\w+'
SPECIAL_CHARS = '&lt;|&lt;|&amp;|#'
PARA='\n+'
def clean(text):
    text = re.sub(LINK, ' ', text)
    text = re.sub(SPECIAL_CHARS, ' ', text)
    text = re.sub(PARA, '\n', text)
    return text

text = clean(text)
text
```

The code generates the following output:

```
'\ufeffThe Project Gutenberg eBook, The Life and Adventures of Robinson Crusoe,\nby Daniel Defoe\nThis eBo
ok is for the use of anyone anywhere in the United States and most\nother parts of the world at no cost an
d with almost no restrictions\nwhatsoever.  You may copy it, give it away or re-use it under the terms of
\nthe Project Gutenberg License included with this eBook or online at\nwww.gutenberg.org.  If you are not
located in the United States, you\'ll have\nto check the laws of the country where you are located before
using this ebook.\nTitle: The Life and Adventures of Robinson Crusoe\nAuthor: Daniel Defoe\nRelease Date:
September 7, 2015  [eBook  521]\nThis file was first posted on February 28, 1996\nLast Updated: November 1
6, 2016\nLanguage: English\nCharacter set encoding: UTF-8\n***START OF THE PROJECT GUTENBERG EBOOK THE LIF
E AND ADVENTURES OF\nROBINSON CRUSOE***\nTranscribed from the 1919 Seeley, Service & Co. edition by David
Price,\nemail ccx074@pglaf.org\n                                              The\n                        Life an
d Adventures\n                                  of\n
By\n                         Daniel Defoe\n                          * * * * *\n
_With Illustrations by H. M. Brock_\n                          * * * * *\n
London\n                     Seeley, Service & Co. Limited\n              38 Great Russell Str
```

Figure 5.18: Text from The Life and Adventures of Robinson Crusoe

4. Insert a new cell and add the following code to import the necessary libraries.
    Clean the text using the spaCy English language model to tokenize the corpus
    and to exclude all tokens that are not detected as nouns:

```
import spacy
nlp = spacy.load('en_core_web_sm')
doc=nlp(text)
"""

We can experiment with other or a combinations of parts of speech
['NOUN','ADJ','VERB','ADV'] #['NOUN','ADJ']
"""

pos_list=['NOUN']


preproc_text=[]
preproc_sent=[]


for token in doc:
    if token.text!='\n':
        if not(token.is_stop) and not(token.is_punct) \
        and token.pos_ in pos_list:
            preproc_sent.append(token.lemma_)
    else:
        preproc_text.append(preproc_sent)
        preproc_sent=[]


preproc_text.append(preproc_sent) #last sentence


print(preproc_text)
```

The code generates the following output:

```
[[], [], ['use'], ['part', 'world', 'cost', 'restriction'], ['term'], [], [], ['law', 'country', 'ebook'],
['title'], ['author'], ['ebook'], ['file'], [], [], [], ['START', 'PROJECT'], [], ['edition'], ['email'],
[], ['year', 'city', 'family'], ['country', 'father', 'foreigner'], ['estate', 'merchandise'], ['trade'],
['mother', 'relation', 'family'], ['country'], ['corruption', 'word'], ['Crusoe', 'companion'], [], ['brot
her', 'lieutenant'], ['regiment', 'foot', 'flander'], ['battle'], ['brother'], ['father', 'mother'], ['so
n', 'family', 'trade', 'head'], ['thought', 'father'], ['share', 'learning'], ['education', 'country', 'sc
hool'], ['law', 'sea'], ['inclination'], ['command', 'father', 'entreaty', 'persuasion'], ['mother', 'frie
nd'], ['propensity', 'nature', 'life', 'misery'], [], ['father', 'man', 'counsel'], ['design', 'morning'],
['chamber', 'gout'], ['subject', 'reason'], ['inclination', 'house'], ['country', 'prospect'], ['fortune',
```

Figure 5.19: Tokenized corpus after preprocessing is done

5. Insert a new cell and add the following code to import the necessary libraries. Create an HDP model with the **α** concentration parameter as **0.8** and see how the negative log likelihood varies with the number of iterations:

```python
import tomotopy as tp


mdl = tp.HDPModel(alpha=0.8,seed=0)


for line in preproc_text:
    mdl.add_doc(line)


for i in range(0, 110, 10):
    mdl.train(i)
    print('Iteration: {}\tLog-likelihood: {}'.\
        format(i, mdl.ll_per_word))


for k in range(mdl.k):
    print('Top 10 words of topic #{}'.format(k))
    print(mdl.get_topic_words(k, top_n=7))
```

The code generates the following output:

```
Iteration: 0    Log-likelihood: -7.90226765159165
Iteration: 10   Log-likelihood: -7.371047840028124
Iteration: 20   Log-likelihood: -7.37652709605804
Iteration: 30   Log-likelihood: -7.428692292874718
Iteration: 40   Log-likelihood: -7.40457823874009
Iteration: 50   Log-likelihood: -7.4007174035615515
Iteration: 60   Log-likelihood: -7.397126200841502
Iteration: 70   Log-likelihood: -7.386784886923981
Iteration: 80   Log-likelihood: -7.3724504423195345
Iteration: 90   Log-likelihood: -7.387830544015653
Iteration: 100  Log-likelihood: -7.375331385114747
```

Figure 5.20: Variation of negative log likelihood with the number of iterations

6. Insert a new cell and add the following code to save the topic model:

```
mdl.save('../data/robinsoncrusoe/hdp_model.bin')
```

7. Insert a new cell and add the following code to load the topic model:

```
mdl = tp.HDPModel.load('../data/robinsoncrusoe/'\
                       'hdp_model.bin')
```

8. Insert a new cell and add the following code to see the probability distribution of topics if you consider the entire dataset as a single document:

```
bag_of_words=[word for sent in preproc_text for word in sent]
doc_inst = mdl.make_doc(bag_of_words)
mdl.infer(doc_inst)[0]
np.argsort(np.array(mdl.infer(doc_inst)[0]))[::-1]
```

The code generates the following output:

```
array([[163, 103,  28,  64, 124,  40, 160, 100, 162, 102, 161, 101,   0,
          1,  49,  48, 169, 168,  72,  73,  35,  47,  43, 107,  45,  69,
         71,  67,  83,  81,  79,  77, 167, 165,  97,  96,  95,   2,  14,
         26,  38,  50,  62,  74,  86,  98, 110, 122, 134, 146, 158, 194,
         18, 114, 138,   6, 183,  32,  92, 184, 186, 119,  11, 113,  17,
          5,  23, 143, 137,  89,  53,  54, 179,  52,  51,  55, 180,  57,
         56,  70, 181,  58,  59,  60,  68,  66,  65,  63,  61,  39, 185,
         46, 192,   3,   7,   8,   9,  12,  13,  15,  19,  20,  21,  24,
         25,  27, 191,  29,  30,  31,  33,  34,  36,  37,  41,  42, 188,
         44, 187, 178, 176,  75, 117,  76, 123, 164, 125, 126, 127, 128,
        129, 130, 131, 132, 133, 135, 139, 140, 141, 144, 145, 147, 148,
        149, 150, 151, 152, 153, 154, 155, 156, 157, 120, 121, 193,  87,
         90,  91,  84, 174,  93,  94, 116,  82, 175, 173, 172, 171, 170,
        104, 105, 106, 166, 108, 109, 111,  80,  78, 115, 177,  85,  99,
        159, 190, 182, 189,  10, 142,  22, 118,  16, 136,   4, 112,  88],
      dtype=int64)
```

Figure 5.21: Probability distribution if the entire corpus is considered

9. Insert a new cell and add the following code to see the probability distribution of topic 163:

```
print(mdl.get_topic_words(163, top_n=7))
```

The code generates the following output:

```
[('horse', 0.13098040223121643), ('way', 0.026405228301882744),
('mankind', 0.26405228301882744), ('fire', 0.026405228301882744),
('object', 0.026405228301882744), ('bridle', 0.026405228301882744),
('distress', 0.026405228301882744)]
```

10. Insert a new cell and add the following code to see the probability distribution of topic 103:

```
print(mdl.get_topic_words(103, top_n=7))
```

The code generates the following output:

```
[('manor', 0.03706422075629234), ('inheritance', 0.03706422075629234),
('lord', 0.03706422075629234), ('man', 0.0003669724682377309),
('shore', 0.0003669724682377309), ('ship',0.0003669724682377309)]
```

11. Insert a new cell and add the following code to see the probability distribution of topic 28:

```
print(mdl.get_topic_words(28, top_n=7))
```

The code generates the following output:

```
[('thought', 0.07716038823127747), ('mind', 0.045609116554260254),
('word', 0.038597721606492996), ('face', 0.03509202599525452),
('terror', 0.03509202599525452), ('tear', 0.3158633038401604),
('apprehension', 0.3158633038401604)]
```

We see that we have 195 topics in all, many of which are likely not interpretable. In general, finding interpretable topics is difficult and connecting the words to interpret topics often requires familiarity with the domain. We have seen that log perplexity has very limited utility. In the case of prior knowledge of the corpus, the topic model has a much smaller role to play in the discovery of the thematic structure.

> **NOTE**
>
> In general, the topics found are extremely sensitive to randomization in both gensim and tomotopy. While setting a `random_state` in gensim could help reproducibility, in general, the topics found using tomotopy are superior from the perspective of interpretability. Generally, your output is expected to be different. In order to have exactly the same topic model, we can save and load topic models, and this was used in this exercise.
>
> To access the source code for this specific section, please refer to https://packt.live/3ggbfAn.
>
> This section does not currently have an online interactive example and will need to be run locally.

We have explored three of the most popular approaches to topic modeling. Let's now discuss the practical challenges in using topic modeling and the state-of-the-art topic modeling technologies.

PRACTICAL CHALLENGES

The selection of the number of topics and topic-modeling algorithms, the number of iterations, and the evaluation of the topic model are the main challenges faced by a practitioner. Having prior knowledge about the domain can greatly help in choosing the number of topics. In the absence of prior knowledge about the expected number of topics, we may need to rely on experimentation for the choice of the topic-modeling algorithm. The HDP model is an attractive choice when there isn't much information about the number of topics. In the case of a small corpus, the LSA model could be used.

One factor that makes interpreting topics difficult is that they contain a lot of very frequently occurring (but indistinctive) words. To overcome this, we can iteratively identify these words and add them to a list of stop words. At times, we may want to filter out words that are too rare and/or too common. The use of only nouns, only verbs, or a combination of various parts of speech can improve the interpretability of topics.

Qualitative evaluation of the topics is essential. We may have to accept a mix of interpretable and non-interpretable topics in the real world. In the absence of human participants, we can use qualitative ways of considering word intrusion. Unless there is a downstream use of the topic model being developed, a complete lack of interpretability will render the topic model useless. When we have a downstream application, even non-interpretable topics are useful as they offer a convenient means to carry out dimensionality reduction on the dataset.

STATE-OF-THE-ART TOPIC MODELING

There is no known benchmark for quantitively identifying the state-of-the-art topic-modeling algorithm. It necessarily involves human participation whenever interpretable topics are required. In cases where the interpretation of topics is not necessary, the topic model needs to be evaluated by downstream tasks. A qualitative approach to interpreting topic models may be useful if there is prior knowledge or familiarity with the corpus.

While there have been attempts at using labeled topic modeling, there is no evidence of these models broadly outperforming unsupervised topic-modeling algorithms. Interestingly, given that much of the topic modeling literature was published prior to 2014, this is not among the most active areas of research. This suggests that complete automation is hard and human participation is here to stay as the state-of-the-art technique in the near future.

ACTIVITY 5.02: COMPARING DIFFERENT TOPIC MODELS

The **Consumer Financial Protection Bureau** (**CFPB**) publishes consumer complaints made against organizations in the financial sector. This original dataset is available at https://www.consumerfinance.gov/data-research/consumer-complaints/#download-the-data. In this activity, you will qualitatively compare how HDP and LDA models perform on the interpretability of topics by analyzing student loan complaints.

> **NOTE**
>
> The dataset to be used for this activity can be found at https://packt.live/39GoyYe.

Follow these steps to complete this activity:

1. Open a Jupyter Notebook.

2. Import the **pandas** library and load the dataset from a text file produced by partially processing the dataset from the CFPB website mentioned at the beginning of this section.

3. Tokenize the text using spaCy. Select tokens that may be a part of speech (noun/verb/adjective or a combination).

4. Train an HDP model.

5. Save and load the HDP model. To save a topic model, use the following line of code:

```
mdl.save('../data/consumercomplaints/hdp_model.bin')
```

To load a topic model, use the following:

```
mdl = tp.HDPModel.load('../data/consumercomplaints/hdp_model.bin')
```

6. Determine the topics in the entire set of complaints. Sample a few topics and check for interpretability.

7. Repeat steps 3-8 for an LDA model instead of an HDP model. Consider the number of topics in the LDA model to around the number of topics found in the HDP model.

8. Select the qualitatively better model from the HDP and LDA models trained in this activity. Also, compare these two models quantitatively.

> **NOTE**
>
> The full solution to this activity can be found on page 400.

In this activity, we successfully compared two different models both qualitatively and quantitatively.

SUMMARY

In this chapter, we discussed topic modeling in detail. Without delving into advanced statistics, we reviewed various topic-modeling algorithms (such as LSA, LDA, and HDP) and how they can be used for topic modeling on a given dataset. We explored the challenges involved in topic modeling, how experimentation can help address those challenges, and, finally, broadly discussed the current state-of-the-art approaches to topic modeling.

In the next chapter, we will learn about vector representation of text, which helps us convert text into a numerical format to make it more easily understandable by machines.

6

VECTOR REPRESENTATION

OVERVIEW

This chapter introduces you to the various ways in which text can be represented in the form of vectors. You will start by learning why this is important, and the different types of vector representation. You will then perform one-hot encoding on words, using the `preprocessing` package provided by scikit-learn, and character-level encoding, both manually and using the powerful Keras library. After covering learned word embeddings and pre-trained embeddings, you will use `Word2Vec` and `Doc2Vec` for vector representation for **Natural Language Processing** (**NLP**) tasks, such as finding the level of similarity between multiple texts.

INTRODUCTION

The previous chapters laid a firm foundation for NLP. But now we will go deeper into a key topic—one that gives us surprising insights into how language processing works and how some of the key advances in human-computer interaction are facilitated. At the heart of NLP is the simple trick of representing text as numbers. This helps software algorithms perform the sophisticated computations that are required to understand the meaning of the text.

As we have already discussed in previous chapters, most machine learning algorithms take numeric data as input and do not understand the text as such. We need to represent our text in numeric form so that we can apply different machine learning algorithms and other NLP techniques to it. These numeric representations are called vectors and are also sometimes called word embeddings or simply embeddings.

This chapter begins with a discussion of vectors, how text can be represented as vectors, and how vectors can be composed to represent complex speech. We will walk through the various representations in both directions—learning how to encode text as vectors as well as how to retrieve text from vectors. We will also look at some cutting-edge techniques used in NLP that are based on the idea of representing text as vectors.

WHAT IS A VECTOR?

The basic mathematical definition of a vector is an object that has both magnitude and direction. In our definition, it is mostly compared with a scalar, which can be defined as an object that has only magnitude. Vectors are also defined as an element in vector space—for example, a point in space with the coordinates (x=4, y=5, z=6) is a vector. Here, we can see the vector dimensions are the geometric coordinates of a point or element in space. However, the vector dimensions can also represent any quantity or property of some element or object in addition to mere geometric coordinates.

As an example, let's say that we're defining the weather at a given place using five features: temperature, humidity, precipitation, wind speed, and air pressure. The units that these would be measured in are Celsius, percentage, centimeters, kilometers per hour (km/h), and millibar (mbar), respectively. The following are the values for two places:

Weather	Place 1	Place 1
Temperature	25	32
Humidity	50	60
Precipitation	1	0
Wind speed	18	7
Air pressure	1200.0	1019.0

Figure 6.1: Weather indicators at two different places

So, we can represent the weather of these places in vector form as follows:

- Vector for place 1: [25, 50, 1, 18, 1200.0]

- Vector for place 2: [32, 60, 0, 7, 1019.0]

In the preceding representation, the first dimension represents temperature, the second dimension represents humidity, and so on. Note that the order of these dimensions should be consistent among all the vectors.

Similarly, we can also represent text as a vector in which each dimension can represent either the presence or absence of certain metrics. Examples of these are bag of words and TFIDF vectors that we looked at in the previous chapters. There are other techniques as well for vector representation of text—learned word embeddings, for instance. We will discuss all these different techniques in the upcoming sections. These techniques can be broadly classified into two categories:

- Frequency-based embeddings

- Learned word embeddings

FREQUENCY-BASED EMBEDDINGS

Frequency-based embedding is a technique in which the text is represented in vector form by considering the frequency of the word in a corpus. The techniques that come under this category are the following:

- Bag of words: As we've already seen in *Chapter 2, Feature Extraction Methods*, bag of words is the technique of converting text into vector or numeric form by representing each sentence or document in a list the length of which is equal to the total number of unique words in the corpus.

- TFIDF: As seen previously in *Chapter 2, Feature Extraction Methods*, this technique considers the frequency of a term as well as the inverse of its occurrence in the corpus.

- Term frequency-based technique: This is a somewhat simpler version of TFIDF. We represent each word in the vector by its number of occurrences in the document. For example, let's say that a document contains the following sentences:

1. The girl is pretty, and the boy is handsome.

2. Do whatever your heart says.

3. The boy has a bike.

4. His bike was red in color.

   Now let's build term frequency vectors of all these sentences. We will first create a **dictionary** of unique words as follows. Note that we are considering every word in lowercase only:

   {1: the

   2: girl

   3: pretty

   4: and

   5: boy

   5: is

   7: handsome

   8: do

   9: whatever

   10: your

   11: heart

   12: says

   13: was

   14: has

   15: bike

   16: his

   17: red

18: in

19: color

}

Now every document will be represented by a vector with 19 dimensions, where every dimension represents the frequency of a word in that document. So, for sentence 1, the vector will be [2, 1, 1, 1, 1, 2, 1, 0, 0, 0, 0, 0, 0, 0, 0, 0, 0, 0, 0]. Similarly, for sentence 2, the vector representation will be [0, 0, 0, 0, 0, 0, 0, 1, 1, 1, 1, 1, 0, 0, 0, 0, 0, 0, 0], and so on. Note that the order needs to be consistent here, too.

> **NOTE**
>
> It is recommended that you use preprocessing techniques such as stemming, stop word removal, and conversion to lowercase before converting a text into the aforementioned vector format. Term frequency is a simple and quick technique for converting text into vector form. However, the TFIDF technique is a more effective technique than term frequency as it not only considers the frequency of a word in the current document but also in the background corpus.

- One-hot encoding: In all techniques described previously, we have represented a word with a single number. Using one-hot encoding, we can represent a word with an array. To understand this concept better, let's take the following sentences:

5. I love cats and dogs.

6. Cats are light in weight.

    We will use a **dictionary** to assign a numeric label or index to each unique word (after converting to lowercase) as follows:

    {1: i

    2: love

    3: cats

    4: and

    5: dogs

6: are

7: light

8: in

9: weight

}

Now we will represent each word in these sentences as follows:

i   [1 0 0 0 0 0 0 0 0]

love   [0 1 0 0 0 0 0 0 0]

cats   [0 0 1 0 0 0 0 0 0]

and   [0 0 0 1 0 0 0 0 0]

dogs [0 0 0 0 1 0 0 0 0]

are   [0 0 0 0 0 1 0 0 0]

light [0 0 0 0 0 0 1 0 0]

in   [0 0 0 0 0 0 0 1 0]

weight [0 0 0 0 0 0 0 0 1]

We can see that each vector consists of 9 elements; that is, the number of elements equals the total number of words in the **dictionary**. For each word, the value of an element will be 1, only if the word is present at the corresponding position in the **dictionary**. When one-hot encoding words, you also need to consider the **vocabulary**. The meaning of vocabulary here is the total number of unique words in the text sources for your project. So, if you have a large source, then you will end up with a huge vocabulary and large one-hot vector sizes, which will eventually consume a lot of memory. The next exercise on word-level one-hot encoding will help us understand this better.

Label encoding is a technique used to convert categorical data in numerical data, where each category is represented by a unique number. In order to perform label encoding and one-hot encoding, we will be using the **LabelEncoder()** and **OneHotEncoder()** classes from the **preprocessing** package provided by the scikit-learn library. The following exercise will help us get a better understanding of this.

EXERCISE 6.01: WORD-LEVEL ONE-HOT ENCODING

In this exercise, we will one-hot encode words with the help of the **preprocessing** package provided by the scikit-learn library. For this, we shall make use of a file containing lines from Jane Austen's *Pride and Prejudice*.

> **NOTE**
>
> The text file used for this exercise can be found at https://packt.live/3hUxNqQ.

Follow these steps to implement this exercise:

1. Open a Jupyter notebook.

2. First, load the file containing the lines from the novel using the **Path** class provided by the **pathlib** library to specify the location of the file. Insert a new cell and add the following code:

```
from pathlib import Path
data = Path('../data')
novel_lines_file = data / 'novel_lines.txt'
```

3. Now that you have the file, open it and read its contents. Use the **open()** and **read()** functions to perform these actions. Store the results in the **novel_lines** file variable. Insert a new cell and add the following code to implement this:

```
with novel_lines_file.open() as f:
    novel_lines_raw = f.read()
```

4. After reading the contents of the file, load it by inserting a new cell and adding the following code:

```
novel_lines_raw
```

The code generates the following output:

```
'It is a truth universally acknowledged, that a single man in possession of a good fortun
e, must be in want of a wife.\nHowever little known the feelings or views of such a man ma
y be on his first entering a neighbourhood, this truth is so well fixed in the minds of th
e surrounding families, that he is considered as the rightful property of some one or othe
r of their daughters.\n"My dear Mr. Bennet," said his lady to him one day, "have you heard
that Netherfield Park is let at last?"\nMr. Bennet replied that he had not.\n"But it is,"
returned she; "for Mrs. Long has just been here, and she told me all about it."\nMr. Benne
t made no answer.\n"Do not you want to know who has taken it?" cried his wife impatientl
y.\n"You want to tell me, and I have no objection to hearing it."\nThis was invitation eno
ugh.\n"Why, my dear, you must know, Mrs. Long says that Netherfield is taken by a young ma
n of large fortune from the north of England; that he came down on Monday in a chaise and
four to see the place, and was so much delighted with it that he agreed with Mr. Morris im
mediately; that he is to take possession before Michaelmas, and some of his servants are t
o be in the house by the end of next week."\n"What is his name?"\n"Bingley."\n"Is he marri
ed or single?"\n"Oh! single, my dear, to be sure! A single man of large fortune; four or f
ive thousand a year. What a fine thing for our girls!"\n"How so? how can it affect the
m?"\n"My dear Mr. Bennet," replied his wife, "how can you be so tiresome! You must know th
at I am thinking of his marrying one of them."\n"Is that his design in settling here?"\n"D
esign! nonsense, how can you talk so! But it is very likely that he may fall in love with
one of them, and therefore you must visit him as soon as he comes."\n"I see no occasion fo
r that. You and the girls may go, or you may send them by themselves, which perhaps will b
e still better, for as you are as handsome as any of them, Mr. Bingley might like you the
best of the party."\n"My dear, you flatter me. I certainly have had my share of beauty, bu
t I do not pretend to be any thing extraordinary now. When a woman has five grown up daugh
ters, she ought to give over thinking of her own beauty."\n"In such cases, a woman has not
often much beauty to think of."\n"But, my dear, you must indeed go and see Mr. Bingley whe
n he comes into the neighbourhood."'
```

Figure 6.2: Raw text from the file

In the output, you will see a lot of newline characters. This is because we loaded the entire content at once into a single variable instead of separate lines. You will also see a lot of non-alphanumeric characters.

5. The main objective is to create one-hot vectors for each word in the file. To do this, construct a vocabulary, which is the entire list of unique words in the file, by tokenizing the string into words and removing newlines and non-alphanumeric characters. Define a function named **clean_tokenize()** to do this. Store the vocabulary created using **clean_tokenize()** inside a variable named **novel_lines**. Add the following code:

```
import string
import re

alpha_characters = str.maketrans('', '', string.punctuation)

def clean_tokenize(text):
    text = text.lower()
    text = re.sub(r'\n', '*** ', text)
    text = text.translate(alpha_characters)
```

```
    text = re.sub(r' +', ' ', text)
    return text.split(' ')


novel_lines = clean_tokenize(novel_lines_raw)
```

6. Take a look at the content inside **novel_lines** now. It should look like a list. Insert a new cell and add the following code to view it:

```
novel_lines
```

The code generates the following output:

```
['it',
 'is',
 'a',
 'truth',
 'universally',
 'acknowledged',
 'that',
 'a',
 'single',
 'man',
 'in',
 'possession',
 'of',
 'a',
 'good',
 'fortune',
 'must',
 'be',
 'in',
```

Figure 6.3: Text after preprocessing is done

7. Insert a new cell and add the following code to convert the list to a NumPy array and print the shape of the array:

```
import numpy as np
novel_lines_array = np.array([novel_lines])
novel_lines_array = novel_lines_array.reshape(-1, 1)
novel_lines_array.shape
```

The code generates the following output:

```
(459, 1)
```

As you can see, the **novel_lines_array** array consists of **459** rows and **1** column. Each row is a word in the original **novel_lines** file.

> **NOTE**
>
> NumPy arrays are more specific to NLP algorithms than Python lists. It is the format that is required for the scikit-learn library, which we will be using to one-hot encode words.

8. Now use encoders, such as the **LabelEncoder()** and **OneHotEncoder()** classes from scikit-learn's **preprocessing** package, to convert **novel_lines_array** to one-hot encoded format. Insert a new cell and add the following lines of code to implement this:

```
from sklearn import preprocessing

labelEncoder = preprocessing.LabelEncoder()
novel_lines_labels = labelEncoder.fit_transform(\
                    novel_lines_array)

import warnings
warnings.filterwarnings('ignore')

wordOneHotEncoder = preprocessing.OneHotEncoder()

line_onehot = wordOneHotEncoder.fit_transform(\
            novel_lines_labels.reshape(-1,1))
```

In the code, the **LabelEncoder()** class encodes the labels, and the **fit_transform()** method fits the label encoder and returns the encoded labels.

9. To check the list of encoded labels, insert a new cell and add the following code:

```
novel_lines_labels
```

The preceding code generates output that looks as follows:

```
array([ 82,  81,   0, 177, 178,   2, 163,   0, 151,  96,  77, 137, 122,
         0,  58,  52, 109,  13,  77, 183, 122,   0, 192,  73,  92,  85,
       164,  45, 127, 181, 122, 156,   0,  96,  99,  13, 125,  70,  47,
        41,   0, 112, 172, 177,  81, 152, 186,  49,  77, 164, 103, 122,
       164, 158,  44, 163,  64,  81,  29,  11, 164, 142, 139, 122, 153,
       126, 127, 128, 122, 165,  31, 110,  33, 106,  17, 143,  70,  86,
       175,  69, 126,  32,  63, 197,  65, 163, 113, 133,  81,  89,  12,
        88, 106,  17, 140, 163,  64,  60, 118,  21,  82,  81, 141, 150,
        51, 107,  93,  62,  83,  15,  68,   7, 150, 176, 100,   5,   1,
        82, 106,  17,  95, 115,   8,  36, 118, 197, 183, 175,  84, 190,
        62, 160,  82,  30,  70, 192,  76, 197, 183, 175, 162, 100,   7,
        74,  63, 115, 120, 175,  66,  82, 172, 184,  80,  40, 191, 110,
        33, 197, 109,  84, 107,  93, 144, 163, 113,  81, 160,  22,   0,
       198,  96, 122,  87,  52,  54, 164, 117, 122,  39, 163,  64,  23,
        37, 125, 104,  77,   0,  27,   7,  53, 175, 145, 164, 136,   7,
       184, 152, 108,  34, 194,  82, 163,  64,   4, 194, 106, 105,  75,
       163,  64,  81, 175, 159, 137,  16, 101,   7, 153, 122,  70, 147,
        10, 175,  13,  77, 164,  71,  22, 164,  38, 122, 114, 185, 187,
        81,  70, 111,  20,  81,  64,  97, 127, 151, 124, 151, 110,  33,
       175,  13, 157,   0, 151,  96, 122,  87,  52,  53, 127,  48, 173,
         0, 196, 187,   0,  46, 169,  51, 130,  55,  72, 152,  72,  24,
        82,   3, 166, 110,  33, 106,  17, 140,  70, 192,  72,  24, 197,
        13, 152, 174, 197, 109,  84, 163,  74,   6, 171, 122,  70,  98,
       126, 122, 166,  81, 163,  70,  35,  77, 148,  68,  35, 116,  72,
        24, 197, 161, 152,  21,  82,  81, 180,  91, 163,  64,  99,  43,
        77,  94, 194, 126, 122, 166,   7, 168, 197, 109, 182,  69,  11,
       154,  11,  64,  28,  74, 145, 115, 121,  51, 163, 197,   7, 164,
        55,  99,  57, 127, 197,  99, 146, 166,  22, 167, 189, 135, 193,
        13, 155,  19,  51,  11, 197,  10,  11,  61,  11,   9, 122, 166,
```

Figure 6.4: List of encoded labels

The **OneHotEncoder()** class encodes the categorical integer features as a one-hot numeric array. The **fit_transform()** method of this class takes the **novel_lines_labels** array as input. This is a numeric array, and each feature included in this array is encoded using the one-hot encoding scheme.

10. Create a binary column for each category. A **sparse matrix** is returned as output. To view the matrix, insert a new cell and type the following code:

```
line_onehot
```

The code generates the following output:

```
<459x199 sparse matrix of type '<class 'numpy.float64'>'
                With 459 stored elements in Compressed Sparse Row
format>
```

11. To convert the sparse matrix into a **dense array**, use the `toarray()` function. Insert a new cell and add the following code to implement this:

```
line_onehot.toarray()
```

The code generates the following output:

```
array([[0., 0., 0., ..., 0., 0., 0.],
       [0., 0., 0., ..., 0., 0., 0.],
       [1., 0., 0., ..., 0., 0., 0.],
       ...,
       [0., 0., 0., ..., 0., 0., 0.],
       [0., 0., 0., ..., 0., 0., 0.],
       [0., 0., 0., ..., 0., 0., 0.]])
```

Figure 6.5: Dense array

> **NOTE**
>
> To access the source code for this specific section, please refer to https://packt.live/2Xd2aAU.
>
> You can also run this example online at https://packt.live/39GSAeu.

The preceding output shows that we have achieved our objective of one-hot encoding words.

One-hot encoding is mostly used in techniques such as language generation models, where a model is trained to predict the next word in the sequence given the words that precede it (think about your phone recommending words while you're chatting with your friends). Language models are used in many important natural language tasks nowadays, including machine translation, spell correction, text summarization, and in tools like Amazon Echo, Alexa, and more.

In addition to word-level language models, we can also build character-level language models, which can be trained to predict the next character in a sequence of characters. For character-level language models, we need character-level one-hot encoding. Let's explore this in the next section.

CHARACTER-LEVEL ONE-HOT ENCODING

In character-level one-hot encoding, we assign a numeric value to all the possible characters. We can use alpha-numeric characters and punctuation as well. Then, we represent each character by an array of size equal to all the characters in the document. This array contains zero at all the indices, other than the index assigned with the character. Let's explain this with an example. Consider the word "hello". Let's say our vocabulary contains only twenty-six characters, so our **dictionary** will look like this:

{'a': 0

'b': 1

'c': 2

'd': 3

'e': 4

'f': 5

'g': 6

'h': 7

'i': 8

'j': 9

'k': 10

.......'z': 25}

Now, 'h' will be represented as [0 0 0 0 0 0 0 1 0 0 0 0 0 0 0 0 0 0 0 0 0 0 0 0 0 0]. Similarly, 'e' can be represented as [0 0 0 0 1 0 0 0 0 0 0 0 0 0 0 0 0 0 0 0 0 0 0 0 0 0]. Let's see how we can implement this in the next exercise.

EXERCISE 6.02: CHARACTER ONE-HOT ENCODING — MANUAL

In this exercise, we will create our own function that can one-hot encode the characters of the word "data". Follow these steps to complete this exercise:

1. Open a Jupyter notebook.

2. To one-hot encode the characters of a given word, create a function named **onehot_word()**. Within this function, create a **lookup** table for each of the characters in the given word. Then, map each character to an index. Add the following code to implement this:

```
def onehot_word(word):
    lookup = {v[1]: v[0] for v in enumerate(set(word))}
    word_vector = []
```

3. Next, loop through the characters in the word and create a vector named **one_hot_vector** of the same size as the number of characters in the lookup. This vector is filled with zeros. Then, use the **lookup** table to find the position of the character and set that character's value to **1**.

> **NOTE**
>
> Execute the code for *step 1* and *step 2* together.

Add the following code:

```
for c in word:
    one_hot_vector = [0] * len(lookup)

    one_hot_vector[lookup[c]] = 1
    word_vector.append(one_hot_vector)
return word_vector
```

The function created earlier will return a word vector.

4. Once the **onehot_word()** function has been created, test it by adding some input as a parameter. Add the word "data" as an input to the function. To implement this, add a new cell and write the following code:

```
onehot_vector = onehot_word('data')
print(onehot_vector)
```

The code generates the following output:

```
[0, 0, 1], [1, 0, 0], [0, 1, 0], [1, 0, 0]
```

Since there are four characters in the input (**data**), there will be four one-hot vectors. To determine the size of each one-hot vector for **data**, we enumerate the total number of characters in it. It is important to note that only one index gets assigned for repeated characters. After enumerating through the characters, the character **d** will be assigned index **0**, the character **a** will be assigned index **1**, and the character **t** will be assigned index **2**.

Based on each character's index position, the elements in each one-hot vector will be marked as **1**, leaving other elements marked **0**. In this way, we can manually one-hot encode any given text. Note that, in most practical applications, the size of one-hot encoded vector is equal to the size of all the characters, and sometimes, non-alphabetical characters are also considered.

> **NOTE**
>
> To access the source code for this specific section, please refer to https://packt.live/314aTX1.
>
> You can also run this example online at https://packt.live/3gaWbE5.

We have learned how character-level one-hot encoding can be performed manually by developing our own function. We will focus on performing character-level one-hot encoding using Keras in the next exercise. Keras is a machine learning library that works along with TensorFlow to create deep learning models.

We will be using the **Tokenizer** class from Keras to create vectors from the text. **Tokenizer** can work on both characters and words, depending on the **char_level** argument. If **char_level** is set to **true**, then it will work on the character level; otherwise, it will work on the word level. The **Tokenizer** class comes with the following functions:

- **fit_on_text()**: This method reads all the text and creates an internal **dictionary**, either word-wise or character-wise. We should always call it for the entire text, so that no word or character is left out of the dictionary. All the methods/variables listed after this should be called or used only after calling this method.

- **word_index**: This is a **dictionary** that contains all the possible words or characters in the vocabulary. Each word or character is assigned a unique number/index.

- **index_word**: This is the reverse dictionary of **word_index**; it contains key-value pairs with the index as the key and the word or character as its value.

- **texts_to_sequences()**: This function converts each word or character sequence into its corresponding index value.

- **texts_to_matrix()**: This converts each word or character in a given text into one-hot vector using a built-in dictionary. It takes the text as input, processes it, and returns a NumPy array of one-hot encoded vectors.

EXERCISE 6.03: CHARACTER-LEVEL ONE-HOT ENCODING WITH KERAS

In this exercise, we will perform one-hot encoding on a given word using the Keras library. Follow these steps to implement this exercise:

1. Open a Jupyter notebook.

2. Insert a new cell and the following code to import the necessary libraries:

```
from keras.preprocessing.text import Tokenizer
import numpy as np
```

3. Once you have imported the **Tokenizer** class, create an instance of it by inserting a new cell and adding the following code:

```
char_tokenizer = Tokenizer(char_level=True)
```

Since you are encoding at the character level, in the constructor, **char_level** is set to **True**.

> **NOTE**
>
> By default, **char_level** is set to **False** if we are encoding words.

4. To test the **Tokenizer** instance, you will require some text to work on. Insert a new cell and add the following code to assign a string to the **text** variable:

```
text = 'The quick brown fox jumped over the lazy dog'
```

5. After getting the text, use the **fit_on_texts()** method provided by the **Tokenizer** class. Insert a new cell and add the following code to implement this:

```
char_tokenizer.fit_on_texts(text)
```

In this code, **char_tokenizer** will break **text** into characters and internally keep track of the tokens, the indices, and everything else needed to perform one-hot encoding.

6. Now, look at the possible output. One type of output is the sequence of the characters—that is, the integers assigned with each character in the text. The **texts_to_sequences()** method of the **Tokenizer** class helps assign integers to each character in the text. Insert a new cell and add the following code to implement this:

```
seq =char_tokenizer.texts_to_sequences(text)
seq
```

The code generates the following output:

```
[[4],
 [5],
 [2],
 [1],
 [9],
 [6],
 [10],
 [11],
 [12],
 [1],
 [13],
 [7],
 [3],
 [14],
 [15],
 [1],
 [16],
 [3],
 [17],
 [1],
 [18],
```

Figure 6.6: List of integers assigned to each character

As you can see, there were **44** characters in the **text** variable. From the output, we can see that for every unique character in **text**, an integer is assigned.

7. Use **sequences_to_texts()** to get text from the sequence with the following code:

```
char_tokenizer.sequences_to_texts(seq)
```

The snippet of the preceding output follows:

```
['t',
 'h',
 'e',
 ' ',
 'q',
 'u',
 'i',
 'c',
 'k',
 ' ',
 'b',
 'r',
 'o',
 'w',
 'n',
 ' ',
 'f',
 'o',
 'x',
 ' ',
 'j',
 'u',
 'm',
 'p',
 'e',
 'd',
 ' ',
 'o',
```

Figure 6.7: Text generated from the sequence

8. Now look at the actual one-hot encoded values. For this, use the **texts_to_matrix()** function. Insert a new cell and add the following code to implement this:

```
char_vectors = char_tokenizer.texts_to_matrix(text)
```

Here, the results of the array are stored in the **char_vectors** variable.

9. In order to view the vector values, just insert a new cell and add the following line:

```
char_vectors
```

On execution, the code displays the array of one-hot encoded vectors:

```
array([[0., 0., 0., ..., 0., 0., 0.],
       [0., 0., 0., ..., 0., 0., 0.],
       [0., 0., 1., ..., 0., 0., 0.],
       ...,
       [0., 0., 0., ..., 0., 0., 0.],
       [0., 0., 0., ..., 0., 0., 0.],
       [0., 0., 0., ..., 0., 0., 1.]])
```

Figure 6.8: Actual one-hot encoded values for the given text

10. In order to investigate the dimensions of the NumPy array, make use of the **shape** attribute. Insert a new cell and add the following code to execute it:

```
char_vectors.shape
```

The following output is generated:

```
(44, 27)
```

So, **char_vectors** is a NumPy array with **44** rows and **27** columns. This is because we are considering 26 characters and an additional character for space.

11. To access the first row of **char_vectors** NumPy array, insert a new cell and add the following code:

```
char_vectors[0]
```

This returns a one-hot vector, which can be seen in the following figure:

```
array([0 ., 0., 0., 0., 1., 0., 0., 0., 0 .,
       0., 0., 0., 0., 0., 0., 0.,0., 0 .,
       0., 0., 0., 0., 0., 0., 0 ., 0., 0])
```

12. To access the index of this one-hot vector, use the **argmax()** function provided by NumPy. Insert a new cell and write the following code to implement this:

```
np.argmax(char_vectors[0])
```

The code generates the following output:

```
4
```

13. The **Tokenizer** class provides two dictionaries, **index_word** and **word_index**, which you can use to view the contents of **Tokenizer** in key-value form. Insert a new cell and add the following code to view the **index_word** dictionary:

```
char_tokenizer.index_word
```

The code generates the following output:

```
{1: ' ',
 2: 'e',
 3: 'o',
 4: 't',
 5: 'h',
 6: 'u',
 7: 'r',
 8: 'd',
 9: 'q',
 10: 'i',
 11: 'c',
 12: 'k',
 13: 'b',
 14: 'w',
 15: 'n',
 16: 'f',
 17: 'x',
 18: 'j',
 19: 'm',
 20: 'p',
 21: 'v',
 22: 'l',
 23: 'a',
 24: 'z',
```

Figure 6.9: The index_word dictionary

As you can see in this figure, the indices act as keys, and the characters act as values. Now insert a new cell and the following code to view the **word_index** dictionary:

```
char_tokenizer.word_index
```

The code generates the following output:

```
{' ': 1,
 'a': 23,
 'b': 13,
 'c': 11,
 'd': 8,
 'e': 2,
 'f': 16,
 'g': 26,
 'h': 5,
 'i': 10,
 'j': 18,
 'k': 12,
 'l': 22,
 'm': 19,
 'n': 15,
 'o': 3,
 'p': 20,
 'q': 9,
 'r': 7,
 't': 4,
 'u': 6,
 'v': 21,
 'w': 14,
 'x': 17,
```

Figure 6.10: The word_index dictionary

In this figure, the characters act as keys, and the indices act as values.

14. In the preceding steps, you saw how to access the index of a given one-hot vector by using the **argmax()** function provided by NumPy. Using this index as a key, you can access its value in the **index_word** dictionary. To implement this, we insert a new cell and write the following code:

```
char_tokenizer.index_word[np.argmax(char_vectors[0])]
```

The preceding code generates the following output:

```
't'
```

In this code, **np.argmax(char_vectors[0])** produces an output of **4**. This will act as a key in finding the value in the **index_word** dictionary. So, when **char_tokenizer.index_word[4]** is executed, it will scan through the dictionary and find that, for key **4**, the value is **t**, and finally, it will print **t**.

> **NOTE**
>
> To access the source code for this specific section, please refer to https://packt.live/2ECjNnf.
>
> You can also run this example online at https://packt.live/2P9c69V.

In the preceding section, we learned how to convert text into one-hot vectors at either the character level or the word level. One-hot encoding is a simple representation of a word, but it has a disadvantage. Whenever the corpus is large (that is, when the number of unique characters or words increases), the size of the one-hot encoded vector also increases. Thus, it becomes very memory intensive and is sometimes not feasible; speed and simplicity here lead to the "curse of dimensionality" by creating a new dimension for each category/word. To tackle this problem, learned embeddings can be used, as explained in the following sections.

LEARNED WORD EMBEDDINGS

The vector representations discussed in the preceding section have some serious disadvantages, as discussed here:

- **Sparsity and large size**: The sizes of one-hot encoded or other frequency-based vectors depend upon the number of unique words in the corpus. This means that when the size of the corpus increases, the number of unique words increases, thereby increasing the size of the vectors in turn.

- **Context**: None of these vector representations consider the words with respect to its context while representing it as a vector. However, the meaning of a word in any language depends upon the context it is used in. Not taking the context into account can often lead to inaccurate results.

Prediction-based word embeddings or learned word embeddings try to address both problems. For starters, these methods represent words with a fixed number of dimensions. Moreover, these representations are actually learned from the different contexts in which the word has been used at different places. **Learned word embeddings** is actually a collective name given to a set of language models that represent words in such a way that words with similar meanings have somewhat similar representations. There are different techniques for creating learned word embeddings, such as `Word2Vec` and GloVe. Let's discuss them one by one.

WORD2VEC

`Word2Vec` is a prediction-based algorithm that represents a word by a vector of a fixed size. This is a form of unsupervised learning algorithm, which means that we need not to provide manually annotated data; we just feed the raw text. It will train a model in such a way that each word is represented in terms of its context throughout the training data.

This algorithm has two variations, as follows:

- **Continuous Bag of Words (CBoW)**: This model tends to predict the probability of a word given the context. The learning problem here is to predict the word given a fixed-window context—that is, a fixed set of continuous words in text.

- **Skip-Gram model**: This model is the reverse of the CBoW model, as it tends to predict the context of a word.

These vectors find application in a lot of NLP tasks including text generation, machine translation, speech to text, text to speech, text classification, and text similarity.

Let's explore how they can be used for text similarity. Suppose we generated 300 dimensional vectors from words such as "love", "adorable", and "hate". If we find the cosine similarity between the vectors for "love" and "adorable", and "love" and "hate", we will find a higher similarity between the former pair of words than the latter.

In the next exercise, we will train word vectors using the gensim library. Specifically, we'll be using the **Word2Vec** class. The **Word2Vec** class has parameters such as **documents, size, window, min_count**, and **workers**. Here, **documents** refers to the sentences that we have to provide to the class, **size** represents the length of the dense vector to represent each token, **min_count** represents the minimum count of words that can be taken into consideration when training a particular model, and **workers** represents the number of threads that are required when training a model.

For training a model, we use the **model.train()** method. This method takes arguments such as **documents, total_examples**, and **epochs**. Here, **documents** represents the sentences, and **total_examples** represents the count of sentences, while **epochs** represents the total number of iterations over the given data. Finally, the trained word vectors get stored in **model.wv**, which is an instance of **KeyedVectors**.

In order to perform basic text cleaning, before it's processed, we will make use of the **textcleaner** class from gensim. Some of the most useful functions available in **textcleaner** that we will be using are as follows:

- **split_sentences()**: As the name suggests, this function splits the text and gets a list of sentences from the text.

- **simple_preprocess()**: This function converts a document into a list consisting of lowercase tokens.

Let's see how we can use these functions to create word vectors.

EXERCISE 6.04: TRAINING WORD VECTORS

In this exercise, we will train word vectors. We will be using books freely available on Project Gutenberg for this. We will also see the vector representation using Matplotlib's pyplot framework.

> **NOTE**
>
> The file we are using for this exercise can be found at https://packt.live/39JeZYP.

Follow these steps to implement this exercise:

1. Open a Jupyter notebook.

2. Use the **requests** library to load books from the Project Gutenberg website, the **json** library to load a book catalog, and the **regex** package to clean the text by removing newline characters. Insert a new cell and add the following code to implement this:

```
import requests
import json
import re
```

3. After importing all the necessary libraries, load the **json** file, which contains details of 10 books, including the title, the author, and the ID. Insert a new cell and add the following steps to implement this:

```
with open('../data/ProjectGutenbergBooks.json', 'r') \
    as catalog_file:
    catalog = json.load(catalog_file)
```

4. To print the details of all the books, insert a new cell and add the following code:

```
catalog
```

The preceding code generates the following output:

```
[{'author': 'Jane Austen', 'id': 1342, 'title': 'Pride and Prejudice'},
 {'author': 'Charles Dickens',
  'id': 46,
  'title': 'A Christmas Carol in Prose'},
 {'author': 'Charles Dickens', 'id': 98, 'title': 'A Tale of Two Cities'},
 {'author': 'Mary Wollstonecraft Shelley',
  'id': 84,
```

Figure 6.11: Book details in the catalog

5. Create a function named **load_book()**, which will take **book_id** as a parameter and, based on that **book_id**, fetch the book and load it. It should also clean the text by removing the newline characters. Insert a new cell and add the following code to implement this:

```
GUTENBERG_URL ='https://www.gutenberg.org/files/{}/{}-0.txt'


def load_book(book_id):
    url = GUTENBERG_URL.format(book_id, book_id)
    contents = requests.get(url).text
    cleaned_contents = re.sub(r'\r\n', ' ', contents)
    return cleaned_contents
```

6. Once you have defined our **load_book()** function, you will loop through the catalog, fetch all the **id** instances of the books, and store them in the **book_ids** list. The **id** instances stored in the **book_ids** list will act as parameters for our **load_book()** function. The book information fetched for each book ID will be loaded in the **books** variable. Insert a new cell and add the following code to implement this:

```
book_ids = [ book['id'] for book in catalog ]
books = [ load_book(id) for id in book_ids]
```

To view the information of the **books** variable, add the following code in a new cell:

```
books[:5]
```

A snippet of the output generated by the preceding code is as follows:

```
["ï»¿ The Project Gutenberg EBook of Pride and Prejudice, by Jane Austen  This eBook is f
or the use of anyone anywhere at no cost and with almost no restrictions whatsoever.  You
may copy it, give it away or re-use it under the terms of the Project Gutenberg License i
ncluded with this eBook or online at www.gutenberg.org   Title: Pride and Prejudice  Auth
or: Jane Austen  Release Date: August 26, 2008 [EBook #1342] Last Updated: November 12, 2
019   Language: English  Character set encoding: UTF-8  *** START OF THIS PROJECT GUTENBE
RG EBOOK PRIDE AND PREJUDICE ***     Produced by Anonymous Volunteers, and David Widger
THERE IS AN ILLUSTRATED EDITION OF THIS TITLE WHICH MAY VIEWED AT EBOOK [# 42671 ]  cover
Pride and Prejudice      By Jane Austen           CONTENTS          Chapter 1
Chapter 2           Chapter 3           Chapter 4           Chapter 5           Chapter 6
Chapter 7           Chapter 8           Chapter 9           Chapter 10          Chapter
11          Chapter 12          Chapter 13          Chapter 14          Chapter 15
Chapter 16          Chapter 17          Chapter 18          Chapter 19          Chapt
er 20          Chapter 21          Chapter 22          Chapter 23          Chapter 24
Chapter 25          Chapter 26          Chapter 27          Chapter 28          Chapt
er 29          Chapter 30          Chapter 31          Chapter 32          Chapter 33
Chapter 34          Chapter 35          Chapter 36          Chapter 37          Chapt
er 38          Chapter 39          Chapter 40          Chapter 41          Chapter 42
Chapter 43          Chapter 44          Chapter 45          Chapter 46          Chapt
```

Figure 6.12: Information of various books

7. Before you can train the word vectors, you need to split the books into a list of documents. In this case, you want to teach the **Word2Vec** algorithm about words in the context of the sentences that they are in. So here, a document is actually a sentence. Thus, you need to create a list of sentences from all 10 books. Insert a new cell and add the following code to implement this:

```
from gensim.summarization import textcleaner
from gensim.utils import simple_preprocess

def to_sentences(book):
    sentences = textcleaner.split_sentences(book)
    sentence_tokens = [simple_preprocess(sentence) \
                    for sentence in sentences]
    return sentence_tokens
```

In the preceding code, all the text preprocessing takes place inside the **to_sentences()** function that you have defined.

8. Now, loop through each book in **books** and pass each book as a parameter to the **to_sentences ()** function. The results should be stored in the **book_sentences** variable. Also, split books into sentences and sentences into documents. The result should be stored in the **documents** variable. Insert a new cell and add the following code to implement this:

```
books_sentences = [to_sentences(book) for book in books]
documents = [sentence for book_sent in books_sentences \
             for sentence in book_sent]
```

9. To check the length of the documents, use the **len ()** function as follows:

```
len(documents)
```

The code generates the following output:

```
32922
```

10. Now that you have your documents, train the model by making use of the **Word2Vec** class provided by the gensim package. Insert a new cell and add the following code to implement this:

```
from gensim.models import Word2Vec
# build vocabulary and train model
model = Word2Vec(
        documents,
        size=100,
        window=10,
        min_count=2,
        workers=10)
model.train(documents, total_examples=len(documents), \
        epochs=50)
```

The code generates the following output:

```
(27809439, 37551450)
```

Now make use of the **most_similar ()** function of the **model.wv** instance to find the similar words. The **most_similar ()** function takes **positive** as a parameter and returns a list of strings that contribute positively. Insert a new cell and add the following code to implement this:

```
model.wv.most_similar(positive="worse")
```

The code generates the following output:

```
[('kinder', 0.6370856761932373),
 ('older', 0.616365909576416),
 ('more', 0.616054892539978),
 ('narrower', 0.6126840710639954),
 ('better', 0.5983243584632874),
 ('larger', 0.5939739346504211),
 ('stronger', 0.5806257128715515),
 ('happier', 0.5734115839004517),
 ('less', 0.5692222714424133),
 ('handsomer', 0.5489497184753418)]
```

Figure 6.13: Most similar words

> **NOTE**
>
> You may get a slightly different output as the output depends on the model training process, so you may have a different model than the one we have trained here.

11. Create a **show_vector()** function that will display the vector using **pyplot**, a plotting framework in Matplotlib. Insert a new cell and add the following code to implement this:

```python
%matplotlib inline
import matplotlib.pyplot as plt

def show_vector(word):
    vector = model.wv[word]
    fig, ax = plt.subplots(1,1, figsize=(10, 2))
    ax.tick_params(axis='both', \
                   which='both',\
                   left=False, \
                   bottom=False, \
                   top=False,\
                   labelleft=False, \
                   labelbottom=False)
```

```
    ax.grid(False)
    print(word)
    ax.bar(range(len(vector)), vector, 0.5)

show_vector('sad')
```

The code generates the following output:

sad

Figure 6.14: Graph of the vector when the input is "sad"

> **NOTE**
>
> To access the source code for this specific section, please refer to https://packt.live/317nb11.
>
> You can also run this example online at https://packt.live/2BJC40I.

In the preceding figure, we can see the vector representation when the word provided to the **show_vector()** function is "sad". We have learned about training word vectors and representing them using **pyplot**. In the next section, we will focus more on using **pre-trained word vectors**, which are required for NLP projects.

USING PRE-TRAINED WORD VECTORS

For a machine learning model, the more data you have, the better the model you get. But training the model on large amounts of data is intensively resource-consuming in terms of both time and memory. So, we usually train a **Word2Vec** model on a large amount of data and retain the model for future use. There are also a lot of pre-trained models publicly available have been trained on huge datasets such as Wikipedia articles. These models include gensim by fastText (research group by Facebook), and **Word2Vec** has recently proved to be state-of-the-art for tasks including checking for word analogies and word similarities, as follows:

- *vector('Paris') - vector('France') + vector('Italy')* results in a vector that is very close to *vector('Rome')*.

- *vector('king') - vector('man') + vector('woman')* is close to *vector('queen')*.

Google's publicly available glove model is similar to the **Word2Vec** model and has produced incredible results. In some applications, we may need to train a **Word2Vec** model on our own specific dataset rather than train a new model from scratch; that is, we can train a pre-trained model on more data. This process is called transfer learning. Transfer learning is based on the concept of transferring knowledge from one domain into another.

> **NOTE**
>
> Pre-trained word vectors can get pretty large. For example, vectors trained on Google News contain 3 million words, and on disk, its compressed size is 1.5 GB.

To better understand how we can use pre-trained word vectors in Python, let's walk through a simple exercise.

EXERCISE 6.05: USING PRE-TRAINED WORD VECTORS

In this exercise, we will load and use pre-trained word embeddings. We will also show the image representation of a few word vectors using the pyplot framework of the Matplotlib library. We will be using **glove6B50d.txt**, which is a pre-trained model.

> **NOTE**
>
> The pre-trained model being used for this file can be found at https://www.kaggle.com/watts2/glove6b50dtxt/download. Download this file and place it in the **data** folder of *Chapter 6, Vector Representation*.

Follow these steps to complete this exercise:

1. Open a Jupyter notebook.

2. Add the following statement to import the **numpy** library:

```
import numpy as np
import zipfile
```

3. Move the downloaded model from the preceding link to the location given in the following code snippet. In order to extract data from a ZIP file, use the **zipfile** Python package. Add the following code to unzip the embeddings from the ZIP file:

```
GLOVE_DIR = '../data/'
GLOVE_ZIP = GLOVE_DIR + 'glove6B50d.txt.zip'
print(GLOVE_ZIP)

zip_ref = zipfile.ZipFile(GLOVE_ZIP, 'r')
zip_ref.extractall(GLOVE_DIR)
zip_ref.close()
```

4. Define a function named **load_glove_vectors()** to return a model Python dictionary. Insert a new cell and add the following code to implement this:

```
def load_glove_vectors(fn):
    print("Loading Glove Model")
    with open( fn,'r', encoding='utf8') as glove_vector_file:
        model = {}
        for line in glove_vector_file:
```

```
            parts = line.split()
            word = parts[0]
            embedding = np.array([float(val) \
                    for val in parts[1:]])
            model[word] = embedding
        print("Loaded {} words".format(len(model)))
    return model

glove_vectors = load_glove_vectors(GLOVE_DIR +'glove6B50d.txt')
```

Here, **`glove_vector_file`** is a text file containing a dictionary. In this, words act as keys and vectors act as values. So, we need to read the file line by line, split it, and then map it to a Python dictionary. The preceding code generates the following output:

```
Loading Glove Model
Loaded 400000 words
```

If we want to view the values of **`glove_vectors`**, then we insert a new cell and add the following code:

```
glove_vectors
```

You will get the following output:

```
{'players': array([-9.1399e-01,  8.4152e-02,  6.4889e-02,  3.8138e-01,  1.0120e-01,
        -4.1253e-01, -1.0374e+00,  5.8694e-01, -1.0811e+00,  4.1565e-01,
         7.5995e-01,  5.8594e-01, -1.0692e+00,  2.8048e-01,  1.0978e+00,
        -2.2174e-02,  5.0837e-01, -1.1568e-02, -7.8902e-01, -1.2340e+00,
        -1.1572e+00,  3.1983e-01,  4.2662e-01,  5.2228e-01, -2.8263e-01,
        -1.1629e+00,  3.8899e-01, -5.7561e-01, -3.0536e-01, -1.0698e+00,
         3.4031e+00,  1.2970e+00,  4.0442e-01, -5.0792e-01,  8.6177e-01,
         7.7060e-01,  4.8023e-01,  4.9316e-01, -4.2102e-01, -8.6115e-01,
        -1.3608e-02,  7.4204e-02,  3.6231e-02,  1.1018e+00,  1.3154e-01,
         2.0627e-01, -2.9658e-03,  6.5953e-01, -7.2998e-01, -1.9931e-01]),
 '1.3277': array([-3.2147e-01, -7.2671e-01,  1.9024e-01, -9.0861e-01, -1.4139e+00,
        -7.5414e-01, -1.0425e+00,  1.2735e-01, -5.3525e-01,  6.2581e-01,
        -9.0084e-01, -5.6066e-01,  1.4616e+00,  2.9404e-01,  6.9728e-01,
        -6.9704e-01, -2.4583e-01, -7.3691e-01, -1.0720e-03,  1.7516e+00,
         1.5047e+00, -3.2935e-02, -5.5113e-01,  1.7669e-01, -1.1550e-01,
         6.8808e-01,  1.2447e+00, -1.2322e+00, -7.0814e-01,  1.5486e-02,
        -1.1982e+00,  3.3936e-01,  2.7709e-01, -5.5773e-01, -7.4283e-01,
         1.0060e+00, -3.6225e-01,  2.9963e-01, -7.9166e-01, -6.2066e-01,
        -1.3871e-01, -1.2618e-01,  8.4840e-01,  2.9157e-01, -8.4433e-01,
         6.1656e-01, -3.3926e-01,  1.1435e-01, -1.3091e-01, -1.3911e+00]),
```

Figure 6.15: Dictionary of glove_vectors

The order of the result dictionary can vary as it is a Python dict.

5. The **glove_vectors** object is basically a dictionary containing the mappings of the words to the vectors, so you can access the vector for a word, which will return a 50-dimensional vector. Insert a new cell and add the code to check the vector for the word **dog**:

```
glove_vectors["dog"]
```

```
array([ 0.11008  , -0.38781  , -0.57615  , -0.27714  ,  0.70521  ,
        0.53994  , -1.0786   , -0.40146  ,  1.1504   , -0.5678   ,
        0.0038977,  0.52878  ,  0.64561  ,  0.47262  ,  0.48549  ,
       -0.18407  ,  0.1801   ,  0.91397  , -1.1979   , -0.5778   ,
       -0.37985  ,  0.33606  ,  0.772    ,  0.75555  ,  0.45506  ,
       -1.7671   , -1.0503   ,  0.42566  ,  0.41893  , -0.68327  ,
        1.5673   ,  0.27685  , -0.61708  ,  0.64638  , -0.076996 ,
        0.37118  ,  0.1308   , -0.45137  ,  0.25398  , -0.74392  ,
       -0.086199 ,  0.24068  , -0.64819  ,  0.83549  ,  1.2502   ,
       -0.51379  ,  0.04224  , -0.88118  ,  0.7158   ,  0.38519  ])
```

Figure 6.16: Array of glove vectors with an input of dog

In order to see the vector for the word **cat**, add the following code:

```
glove_vectors["cat"]
```

```
array([ 0.45281 , -0.50108 , -0.53714 , -0.015697,  0.22191 ,  0.54602
       ,
       -0.67301 , -0.6891  ,  0.63493 , -0.19726 ,  0.33685 ,  0.7735
       ,
        0.90094 ,  0.38488 ,  0.38367 ,  0.2657  , -0.08057 ,  0.61089
       ,
       -1.2894  , -0.22313 , -0.61578 ,  0.21697 ,  0.35614 ,  0.44499
       ,
        0.60885 , -1.1633  , -1.1579  ,  0.36118 ,  0.10466 , -0.78325
       ,
        1.4352  ,  0.18629 , -0.26112 ,  0.83275 , -0.23123 ,  0.32481
       ,
        0.14485 , -0.44552 ,  0.33497 , -0.95946 , -0.097479,  0.48138
       ,
       -0.43352 ,  0.69455 ,  0.91043 , -0.28173 ,  0.41637 , -1.2609
       ,
        0.71278 ,  0.23782 ])
```

Figure 6.17: Array of glove vectors with an input of cat

6. Now that you have the vectors, represent them as an image using the pyplot framework of the Matplotlib library. Insert a new cell and add the following code to implement this:

```
%matplotlib inline
import matplotlib.pyplot as plt


def to_vector(glove_vectors, word):
    vector = glove_vectors.get(word.lower())
    if vector is None:
        vector = [0] * 50
    return vector


def to_image(vector, word=''):
    fig, ax = plt.subplots(1,1)
    ax.tick_params(axis='both', which='both',\
                   left=False, \
                   bottom=False, \
                   top=False,\
                   labelleft=False,\
                   labelbottom=False)
    ax.grid(False)
    ax.bar(range(len(vector)), vector, 0.5)
    ax.text(s=word, x=1, y=vector.max()+0.5)
    return vector
```

In the preceding code, you defined two functions. The **to_vector()** function accepts **glove_vectors** and **word** as parameters. Here, the **get()** function of **glove_vectors** will find the word and convert it into lowercase. The result will be stored in the **vector** variable.

7. The **to_image()** function takes **vector** and **word** as input and shows the image representation of **vector**. To find the image representation of the word **man**, type the following code:

```
man = to_image(to_vector(glove_vectors, "man"))
```

The code generates the following output:

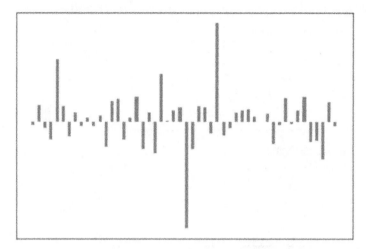

Figure 6.18: Graph generated with an input of man

8. To find the image representation of the word **woman**, type the following code:

```
woman = to_image(to_vector(glove_vectors, "woman"))
```

This will generate the following output:

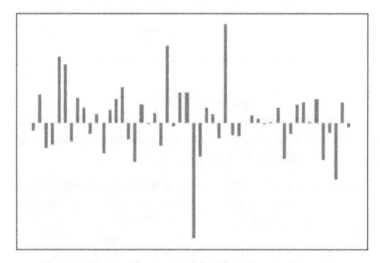

Figure 6.19: Graph generated with an input of woman

9. To find the image representation of the word **king**, type the following code:

```
king = to_image(to_vector(glove_vectors, "king"))
```

This will generate the following output:

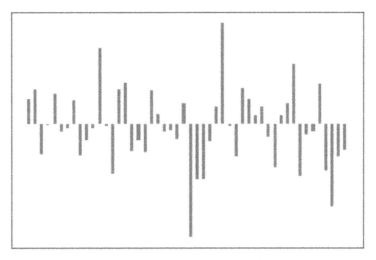

Figure 6.20: Graph generated with an input of king

10. To find the image representation of the word **queen**, type the following code:

```
queen = to_image(to_vector(glove_vectors, "queen"))
```

This will generate the following output:

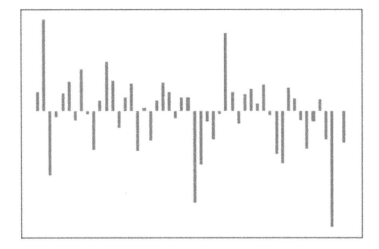

Figure 6.21: Graph generated with an input of queen

11. To find the image representation of the vector for **king − man + woman − queen**, type the following code:

```
diff = to_image(king - man + woman - queen)
```

This will generate the following output:

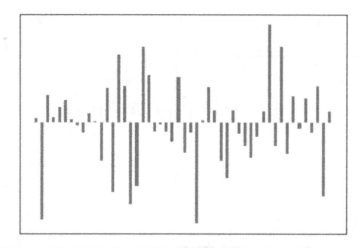

Figure 6.22: Graph generated with (king-man+woman-queen) as input

12. To find the image representation of the vector for **king − man + woman**, type the following code:

```
nd = to_image(king - man + woman)
```

This will generate the following output:

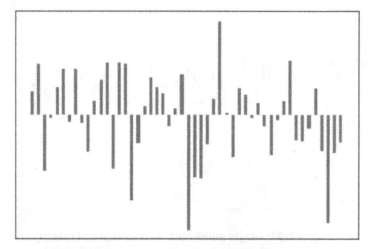

Figure 6.23: Graph generated with (king-man+woman) as input

> **NOTE**
>
> To access the source code for this specific section, please refer to https://packt.live/33btLpH.
>
> This section does not currently have an online interactive example, and will need to be run locally.

The preceding results are the visual proof of the example we already discussed. We've learned how to load and use pre-trained word vectors and view their image representations. In the next section, we will focus on document vectors and their uses.

DOCUMENT VECTORS

Word vectors and word embeddings represent words. But if we wanted to represent a whole document, we'd need to use document vectors. Note that when we refer to a document, we are referring to a collection of words that have some meaning to a user. A document can be a single sentence or a group of sentences. A document can consist of product reviews, tweets, or lines of movie dialogue, and can be from a few words to thousands of words. A document can be used in a machine learning project as an instance of something that the algorithm can learn from. We can represent a document with different techniques:

- Calculating the mean value: We calculate the mean of all the constituent word vectors of a document and represent the document by the mean vector.

- **Doc2Vec**: **Doc2Vec** is a technique by which we represent documents by a fixed-length vector. It is trained quite similarly to the way we train the **Word2Vec** model. Here, we also add the unique ID of the document to which the word belongs. Then, we can get the vector of the document from the trained model using the document ID.

Similar to **Word2Vec**, the **Doc2Vec** class contains parameters such as **min_count**, **window**, **vector_size**, **sample**, **negative**, and **workers**. The **min_count** parameter ignores all the words with a frequency less than that specified. The **window** parameter sets the maximum distance between the current and predicted words in the given sentence. The **vector_size** parameter sets the dimensions of each vector.

The **sample** parameter defines the threshold that allows us to configure the higher-frequency words that are regularly down-sampled, while **negative** specifies the total amount of noise words that should be drawn and **workers** specifies the total number of threads required to train the model. To build the vocabulary from the sequence of sentences, **Doc2Vec** provides the **build_vocab** method. We'll be using all of these in the upcoming exercise.

USES OF DOCUMENT VECTORS

Some of the uses of document vectors are as follows:

- **Similarity**: We can use document vectors to compare texts for similarity. For example, legal AI software can use document vectors to find similar legal cases.

- **Recommendations**: For example, online magazines can recommend similar articles based on those that users have already read.

- **Predictions**: Document vectors can be used as input into machine learning algorithms to build predictive models.

In the next section, we will perform an exercise based on document vectors.

EXERCISE 6.06: CONVERTING NEWS HEADLINES TO DOCUMENT VECTORS

In this exercise, we will convert some news headlines into document vectors. Also, we will look at the image representation of the vector. Again, for image representation, we will be using the pyplot framework of the Matplotlib library. Follow these steps to complete this exercise:

> **NOTE**
>
> The file which we are going to use in this exercise is in zipped format and can be found at https://packt.live/3fhE2TG. It should be unzipped once downloaded.

1. Open a Jupyter notebook.

2. Import all the necessary libraries for this exercise. You will be using the gensim library. Insert a new cell and add the following code:

```
import pandas as pd
from gensim import utils
from gensim.models.doc2vec import TaggedDocument
```

```
from gensim.models import Doc2Vec
from gensim.parsing.preprocessing \
import preprocess_string, remove_stopwords
import random
import warnings
warnings.filterwarnings("ignore")
```

In the preceding code snippet, other than other imports, you imported **TaggedDocument** from gensim, which prepares the document formats used in **Doc2Vec**. It represents the document along with the tag. This will be clearer from the following code lines. **Doc2Vec** requires each instance to be a **TaggedDocument** instance.

3. Move the downloaded file to the following location and create a variable of the path as follows:

```
sample_news_data = '../data/sample_news_data.txt'
```

4. Now load the file:

```
with open(sample_news_data, encoding="utf8", \
        errors='ignore') as f:
    news_lines = [line for line in f.readlines()]
```

5. Now create a DataFrame out of the headlines as follows:

```
lines_df = pd.DataFrame()

indices  = list(range(len(news_lines)))

lines_df['news'] = news_lines
lines_df['index'] = indices
```

6. View the head of the DataFrame using the following code:

```
lines_df.head()
```

This will create the following output:

	news	index
0	Top of the Pops leaves BBC One The BBC flagshi...	0
1	Oscars race enters final furlong The race for ...	1
2	US TV special for tsunami relief A US televisi...	2
3	Williamson lauds bowlers for adapting to atypi...	3
4	Housewives lift Channel ratings The debut of U...	4

Figure 6.24: Head of the DataFrame

7. Create a class, the object of which will create the training instances for the **Doc2Vec** model. Insert a new cell and add the following code to implement this:

```
class DocumentDataset(object):

    def __init__(self, data:pd.DataFrame, column):
        document = data[column].apply(self.preprocess)
        self.documents = [ TaggedDocument( text, [index]) \
                        for index, text in \
                        document.iteritems() ]

    def preprocess(self, document):
        return preprocess_string(\
            remove_stopwords(document))

    def __iter__(self):
        for document in self.documents:
            yield documents

    def tagged_documents(self, shuffle=False):
        if shuffle:
            random.shuffle(self.documents)
        return self.documents
```

In the code, the **preprocess_string()** function applies the given filters to the input. As its name suggests, the **remove_stopwords()** function is used to remove **stopwords** from the given document. Since **Doc2Vec** requires each instance to be a **TaggedDocument** instance, we create a list of **TaggedDocument** instances for each headline in the file.

8. Create an object of the **DocumentDataset** class. It takes two parameters. One is the **lines_df_small** DataFrame and the other is the **Line** column name. Insert a new cell and add the following code to implement this:

```
documents_dataset = DocumentDataset(lines_df, 'news')
```

9. Create a **Doc2Vec** model using the **Doc2Vec** class. Insert a new cell and add the following code to implement this:

```
docVecModel = Doc2Vec(min_count=1, window=5, vector_size=100, \
                      sample=1e-4, negative=5, workers=8)
docVecModel.build_vocab(documents_dataset.tagged_documents())
```

10. Now you need to train the model using the **train()** function of the **Doc2Vec** class. This could take a while, depending on how many records we train. Here, **epochs** represents the total number of records required to train the document. Insert a new cell and add the following code to implement this:

```
docVecModel.train(documents_dataset.\
                  tagged_documents(shuffle=True),\
                  total_examples = docVecModel.corpus_count,\
                  epochs=10)
```

11. Save this model for future use as follows:

```
docVecModel.save('../data/docVecModel.d2v')
```

12. The model has been trained. To verify this, access one of the vectors with its index. To do this, insert a new cell and add the following code to find the **doc** vector of index **657**:

```
docVecModel[657]
```

You should get an output similar to the one below:

```
array([-0.20929255, -0.28592077,  0.07170248, -0.08378136,  0.08677434,
       -0.08534106, -0.10685636,  0.00576654, -0.03399038,  0.31707773,
       -0.01177737, -0.01255067,  0.08325162, -0.00340701,  0.18105389,
        0.21334894, -0.06321663,  0.12173653,  0.06913093,  0.26549008,
       -0.13100868, -0.18686569,  0.15937229, -0.2514969 , -0.04796949,
       -0.16842327,  0.00436876, -0.33365294, -0.28871712,  0.03756697,
        0.04344342,  0.00543962, -0.04138061,  0.18812971, -0.09889945,
        0.04105975, -0.10300889,  0.23657063,  0.13254939, -0.00466744,
        0.11880156,  0.5763065 , -0.13349594,  0.0199004 , -0.20860918,
        0.25577313,  0.19562072, -0.07257853, -0.06559091,  0.10523199,
        0.02009419,  0.16713172, -0.01012599,  0.1330115 ,  0.01896849,
        0.17950307, -0.17225678,  0.12428728,  0.29377568, -0.07040878,
       -0.02737776, -0.08043538,  0.14898372, -0.13726163,  0.19797257,
        0.21860234, -0.09122779, -0.20461401,  0.11374234,  0.26827952,
        0.20443173, -0.06511806, -0.08415387,  0.03433008,  0.12155177,
       -0.2339348 ,  0.13126448,  0.25857013,  0.1554307 ,  0.07587516,
        0.07074967,  0.20744652, -0.1130043 , -0.3087442 , -0.13767944,
        0.02682412, -0.07112097,  0.15238564, -0.05019746, -0.12936442,
        0.12461095, -0.04515083, -0.11713583, -0.11419832, -0.14024146,
       -0.00401924,  0.04734296,  0.03295911, -0.11777124,  0.04519369],
      dtype=float32)
```

Figure 6.25: Lines represented as vectors

13. To check the image representation of any given vector, make use of the pyplot framework of the Matplotlib library. The **show_news_lines()** function takes a line number as a parameter. Based on this line number, find the vector and store it in the **doc_vector** varlable. The **show_image()** function takes two parameters, **vector** and **line**, and displays an image representation of the vector. Insert a new cell and add the following code to implement this:

```
import matplotlib.pyplot as plt

def show_image(vector, line):
    fig, ax = plt.subplots(1,1, figsize=(10, 2))
    ax.tick_params(axis='both', \
                   which='both',\
                   left=False, \
                   bottom=False,\
                   top=False,\
                   labelleft=False,\
                   labelbottom=False)
    ax.grid(False)
    print(line)
```

```
        ax.bar(range(len(vector)), vector, 0.5)



def show_news_lines(line_number):
    line = lines_df[lines_df.index==line_number].news
    doc_vector = docVecModel[line_number]
    show_image(doc_vector, line)
```

14. Now that you have defined the functions, implement the
    show_news_lines() function to view the image representation of
    the vector. Insert a new cell and add the following code to implement this:

```
show_news_lines(872)
```

The code generates the following output:

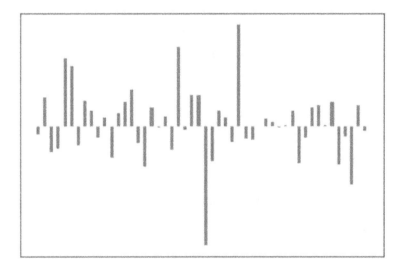

Figure 6.26: Image representation of a given vector

> **NOTE**
>
> To access the source code for this specific section, please refer
> to https://packt.live/30dFxxV.
>
> You can also run this example online at https://packt.live/39MiTQG.

We have learned how to represent a document as a vector. We have also seen a
visual representation of this. In the next section, we will complete an activity to find
similar news headlines using the document vector.

ACTIVITY 6.01: FINDING SIMILAR NEWS ARTICLE USING DOCUMENT VECTORS

To complete this activity, you need to build a news search engine that finds similar news articles like the one provided as input using the **Doc2Vec** model. You will find headlines similar to "US raise TV indecency US politicians are proposing a tough new law aimed at cracking down on indecency." Follow these steps to complete this activity:

1. Open a Jupyter notebook and import the necessary libraries.

2. Load the new article lines file.

3. Iterate over each headline and split the columns and create a DataFrame.

4. Load the **Doc2Vec** model that you created in the previous exercise.

5. Create a function that converts the sentences into vectors and another that does the similarity checks.

6. Test both the functions.

> **NOTE**
>
> The full solution to this activity can be found on page 406.

So, in this activity, we were able to find similar news headlines with the help of document vectors. A common use case of inferring text similarity from document vectors is in text paraphrasing, which we'll explore in detail in the next chapter.

SUMMARY

In this chapter, we learned about the motivations behind converting human language in the form of text into vectors. This helps machine learning algorithms to execute mathematical functions on the text, detect patterns in language, and gain an understanding of the meaning of the text. We also saw different types of vector representation techniques, such as character-level encoding and one-hot encoding.

In the next chapter, we will look at the areas of text paraphrasing, summarization, and generation. We will see how we can automate the process of text summarization using the NLP techniques we have learned so far.

7

TEXT GENERATION AND SUMMARIZATION

OVERVIEW

This chapter begins with the concept of text generation using Markov chains, before moving on to two types of text summarization—namely, abstractive and extractive summarization. You will then explore the TextRank algorithm and use it with different datasets. By the end of this chapter, you will understand the applications and challenges of text generation and summarization using **Natural Language Processing (NLP)** approaches.

INTRODUCTION

The ability to express thoughts in words (sentence generation), the ability to replace a piece of text with different but equivalent text (paraphrasing), and the ability to find the most important parts of a piece of text (summarization) are all key elements of using language. Although sentence generation, paraphrasing, and summarization are challenging tasks in NLP, there have been great strides recently that have made them considerably more accessible. In this chapter, we explore them in detail and see how we can implement them in Python.

GENERATING TEXT WITH MARKOV CHAINS

An idea is expressed using the words of a language. As ideas are not tangible, it is useful to look at text generation in order to gauge whether a machine can think on its own. The utility of text generation is currently limited to an auto-complete functionality, besides a few negative use cases that we will discuss later in this section. Text can be generated in many different ways, which we will explore using Markov chains. Whether this generated text can correspond to a coherent line of thought is something that we will address later in this section.

MARKOV CHAINS

A state space defines all possible states that can exist. A Markov chain consists of a state space and a specific type of successor function. For example, in the case of the simplified state space to describe the weather, the states could be Sunny, Cloudy, or Rainy. The successor function describes how a system in its current state can move to a different state or even continue in the same state. To better understand this, consider the following diagram:

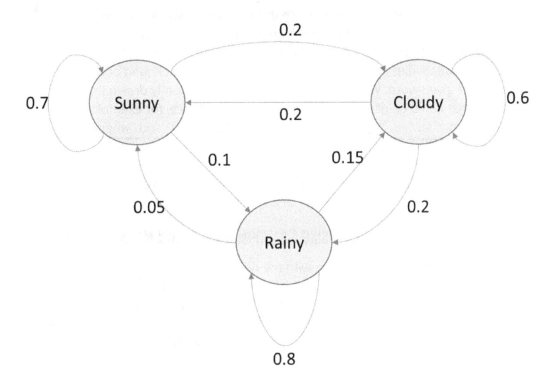

Figure 7.1: Markov chain for weather

The successor function of a Markov chain is a random selection of a successor state based on probabilities. For instance, consider that the initial state is randomly selected as Rainy. The next state could be Rainy (there is a 0.8 probability that the state stays Rainy). Then, the next state could be Sunny (there is a 0.05 probability associated with this transition). It could be Rainy again, and then it could be Cloudy, and so on. Our sequence of states is **Rainy-Rainy-Sunny-Rainy-Cloudy**. For each state, the successor state is found by a random selection; this is called a random walk on the Markov chain.

Similarly, if we have a state space in which the states correspond to a vocabulary, then a random walk on such a Markov chain will generate text. Now, the vocabulary could have around 20,000 words. In this case, the Markov chain will have 20,000 states. The probabilities in this case will correspond to the likelihood of a word succeeding a given word. We can begin with any state randomly drawn from among the words that could be used for the first word of a sentence, for example, common words such as "the, " "a, " "I, " "he, " "she, " "if, " "this, " "why, " and "where". We then find its successor state in a random way, followed by the next successor state found in a random way, and continue in the same manner until we have generated a sequence of words of the required length. In the next section, we will do an exercise related to Markov chains to get a better understanding of them.

EXERCISE 7.01: TEXT GENERATION USING A RANDOM WALK OVER A MARKOV CHAIN

In this exercise, we will generate text with the help of Markov chains. We will use Robert Frost's collection of poems, *North of Boston*, available from Project Gutenberg, to specify the successor state(s) for each state using a dictionary. We'll use a list to specify the successor state(s) for any state so that the number of times a successor state occurs in that list is directly proportional to the probability of transitioning to that successor state.

Then, we will generate 10 phrases with three words in addition to an initial word, and then generate another 10 phrases with four words in addition to an initial word. The initial state or initial word will be randomly selected from among these words: "the," "a," "I," "he," "she," "If," "this," "why," and "where." Note that since we are generating text using a random walk over a Markov chain, in general, the output you get will be different from the output shown in this exercise. Each different output corresponds to new text generation.

> **NOTE**
>
> You can find the text file that's been used for this exercise at https://packt.live/2DiGAE3.

Follow these steps to complete this exercise:

1. Open a Jupyter notebook.

2. Insert a new cell and add the following code to import the necessary libraries and read the dataset:

```
import re
import random
OPEN_DATA_URL = '../data/robertfrost/pg3026.txt'

f=open(OPEN_DATA_URL,'r',encoding='utf-8')
text=f.read()
f.close()
```

3. Insert a new cell and add the following code to preprocess the text using regular expressions:

```
HANDLE = '@\w+\n'
LINK = 'https?://t\.co/\w+'
SPECIAL_CHARS = '&lt;|&lt;|&amp;|#'
PARA='\n+'
def clean(text):
    #text = re.sub(HANDLE, ' ', text)
    text = re.sub(LINK, ' ', text)
    text = re.sub(SPECIAL_CHARS, ' ', text)
    text = re.sub(PARA, '\n', text)
    return text

text = clean(text)
```

4. Split the corpus into a list of words. Show the number of words in the corpus:

```
corpus=text.split()
corpus_length=len(corpus)
corpus_length
```

The preceding code generates the following output:

```
19985
```

5. Insert a new cell and add the following code to define the successor states for each state. Use a dictionary for this:

```
succ_func={}
corpus_counter=0
for token in corpus:
    corpus_counter=corpus_counter+1
    if corpus_counter<corpus_length:
        if token not in succ_func.keys():
            succ_func[token]=[corpus[corpus_counter]]
        else:
            succ_func[token].append(corpus[corpus_counter])
succ_func
```

The preceding code generates an output as follows. Note that we're only displaying a part of the output here.

```
{'\ufeffThe': ['Project'],
 'Project': ['Gutenberg',
  'Gutenberg',
  'Gutenberg',
  'Gutenberg-tm',
  'Gutenberg',
  'Gutenberg-tm',
  'Gutenberg-tm',
  'Gutenberg-tm',
```

Figure 7.2: Dictionary of successor states

We find that "he" is shown as a successor of "who" more than once. This is because this occurs more than once in the dataset. In effect, the number of times the successors occur in the list is proportional to their respective probabilities. Though it is not the only method, this is a convenient way to represent the successor function.

6. Define the list of initial states. Then, define a function to select a random initial state from these and concatenate it with successor states. These successor states are randomly selected from the list containing successor states for a specific current state. Add the following code to do this:

```
initial_states=['The','A','I','He','She','If',\
                'This','Why','Where']
def generate_words(k=5):
```

```
    initial_state=random.choice(initial_states)
    current_state=initial_state
    text=current_state+' '
    for i in range(k):
        succ_state=random.choice(succ_func[current_state])
        text=text+succ_state+' '
        current_state=succ_state
    print(text.split('.')[0])
```

7. Insert a new cell and add the following code to generate text containing 10 phrases of four words (including the initial word) and 10 phrases of five words (including the initial word):

```
for k in range(3,5):
    for j in range(10):
        generate_words(k)
```

The preceding code generates the following output:

```
A horse's hoof pawed
She let me go!"
This and hear some
If you must include
A man's affairs
The rumbling voice said
The lawyer said
The Purple Lady's Slipper
Why didn't see at
If he's kinder than
She does she couldn't lead
He has gone
Where you couldn't keep in
This is your efforts and
The Seven Caves that was
She reached the neighbours, Being
If you 'AS-IS' WITH NO
Why not so With white-faced
I said there and any
Why didn't think we know
```

Figure 7.3: Phrases generated, consisting of four and five words

NOTE

To access the source code for this specific section, please refer to https://packt.live/313fiJY.

You can also run this example online at https://packt.live/33ilO2l.

It's quite interesting that we are able to generate text using a random walk over a Markov chain. If we look more closely, we will see that only a few of the phrases make sense. Broadly speaking, we are generating text that has an element of Robert Frost's style. However, it can hardly be said to correspond to a thought of any kind.

The practical utility of generating text using a Markov chain is somewhat limited to generating spam (spam generators could use a Markov chain) and generating something that is a little amusing. Nevertheless, this exercise demonstrates the surprising results we can get by using a simple approach in which nothing about the structure of a language is explicitly taught to the machine.

In general, auto-complete is one positive use case and arguably the sole positive use case for text generation given that other use cases (besides spam) tend to include the generation of misinformation.

Paraphrasing involves replacing some text with different text that has the same meaning. Now, intuitively, a machine will be able to tell whether one piece of text is a paraphrase of another, but only if that machine understands the meaning. So, one way of checking whether a machine understands the meaning of a piece of text is to check if it can tell if another different piece of text is a paraphrase of that first text.

Benchmark datasets provide a standard touchstone for evaluating approaches to solve a problem. The approaches are typically ranked in a publicly available leaderboard. Even in the case of such benchmark datasets, as of February 21, 2020, the SuperGLUE leaderboard (https://super.gluebenchmark.com/leaderboard) sets human baselines at the top when considered across a variety of tasks. This means that humans are superior at paraphrasing than the most sophisticated approaches even on the specified datasets. Paraphrasing is even tougher outside of benchmark datasets because it is tougher to teach models in a more general way so that the model is as effective for other datasets. Thus, compared to machines, humans can paraphrase even better on other datasets than machines can. In short, paraphrasing using NLP is challenging and is currently of limited practical utility to the practitioner. In the next section, we will learn about summarization.

TEXT SUMMARIZATION

Automated text summarization is the process of using NLP tools to produce concise versions of text that preserve the key information present in the original content. Good summaries can communicate the content with less text by retaining the key information while filtering out other information and noise (or useless text, if any). A shorter text may often take less time to read, and thus summarization facilitates more efficient use of time.

The type of summarization that we are typically taught in school is abstractive summarization. One way to think of this is to consider abstractive summarization as a combination of understanding the meaning and expressing it in fewer sentences. It is usually considered as a supervised learning problem as the original text and the summary are both required. However, a piece of text can be summarized in more than one way. This makes it hard to teach the machine in a general way. While abstractive summarization is an active area of research, it is, for the time being, not at a stage that will be of interest to the practitioner.

There is another form of summarization, called extractive summarization, in which parts of the text are extracted to form a summary. There is no paraphrasing in this form of summarization. This second type will be the focus of the remainder of this section. We will look at the TextRank algorithm, which is an unsupervised machine learning method. For simplicity, we will focus on single-document summarization in this chapter. To implement this, we will be using the gensim library.

TEXTRANK

TextRank is a graph-based algorithm (developed by Rada Mihalcea and Paul Tarau) used to find the key sentences in a piece of text. As we already know, in graph theory, a graph has nodes and edges. In the TextRank algorithm, we estimate the importance of each sentence and create a summary with the sentences that have the highest importance.

The TextRank algorithm works as follows:

1. Represent a unit of text (say, a sentence) as a node.

2. Each node is given an arbitrary importance score.

3. Each edge has a weight that corresponds to the similarity between two nodes (for instance, the sentences **Sx** and **Sy**). The weight could be the number of common words (say, w_k) in the two sentences divided by the sum of the number of words in the two sentences. This can be represented as follows:

$$Similarity\left(S_x, S_y\right) = \frac{\left|\left(W_k : W_k \in Sx \ \& \ W_k \in Sy\right)\right|}{log\left|Sx\right| + log\left|Sy\right|}$$

Figure 7.4: Formula for similarity between two sentences

4. For each node, we compute a new importance score, which is a function of the importance score of the neighboring nodes and the edge weights (w_{ji}) between them. Specifically, the function (f) could be the edge-weighted average score of all the neighboring nodes that are directed toward that node that is adjusted by all the outward edge weights (w_{jk}) and the damping factor (d). This can be represented as follows:

$$f(V_i) = (1 - d) + d \ * \ \sum_{V_j \in In(V_i)} \frac{W_{ji}}{\sum_{V_k \in Out(V_j)} W_{jk}} f(V_j)$$

Figure 7.5: Formula for importance score

d=0.85 is typically used as the damping factor. While we have used a directed graph here, an undirected graph could also be used with a TextRank algorithm.

5. We repeat the preceding step until the importance score varies by less than a pre-defined tolerance level in two consecutive iterations.

6. Sort the nodes in decreasing order of the importance scores.

7. The top n nodes give us a summary.

The number of iterations required for convergence depends on the number of nodes and the connectedness among the nodes. The number of iterations required for an undirected graph is expected to be higher than the number of iterations required for a directed graph since the edges don't have a direction in the case of the former. We typically use a directed graph in the TextRank algorithm. In general, around 20-40 iterations may be required for convergence. We can drop edges that have less than a certain threshold weight for faster convergence since they won't have much of an impact on the result anyway. The basic concept underpinning the TextRank algorithm is that key parts of a document are connected to form a coherent summary.

KEY INPUT PARAMETERS FOR TEXTRANK

We'll be using the gensim library to implement TextRank. The following are the parameters required for this:

- **text**: This is the input text.

- **ratio**: This is the required ratio of the number of sentences in the summary to the number of sentences in the input text.

The gensim implementation of the TextRank algorithm uses BM25—a probabilistic variation of TF-IDF—for similarity computation in place of the similarity measure described in *step 3* of the algorithm. This will be clearer in the following exercise, in which you will summarize text using TextRank.

EXERCISE 7.02: PERFORMING SUMMARIZATION USING TEXTRANK

In this exercise, we will use the classic short story, *After Twenty Years* by O. Henry, which is available on Project Gutenberg, and the first section of the Wikipedia article on Oscar Wilde. We will summarize each text separately so that we have 20% of the sentences in the original text and then have 25% of the sentences in the original text using the gensim implementation of the TextRank algorithm. In all, we shall extract and print four summaries.

In addition to these libraries, you will need to import the following:

```
from gensim.summarization import summarize
summarize(text,ratio=0.20)
```

In the preceding code snippet, **ratio=0.20** means that 20% of the sentences from the original text will be used to create the summary.

> **NOTE**
>
> The text corpus for O. Henry's short story, After Twenty Years, being used in this exercise can be found at https://packt.live/33atvr0.
>
> The Oscar Wilde section from the Wikipedia article can be found at https://packt.live/3fhEocY.

Complete the following steps to implement this exercise:

1. Open a Jupyter notebook.

2. Insert a new cell and add the following code to import the necessary libraries and extract the required text from *After Twenty Years*:

```
from gensim.summarization import summarize
import wikipedia
import re

file_url_after_twenty=r'../data/ohenry/pg2776.txt'
with open(file_url_after_twenty, 'r') as f:
        contents = f.read()

start_string='AFTER TWENTY YEARS\n\n\n'
end_string='\n\n\n\n\n\nLOST ON DRESS PARADE'
text_after_twenty=contents[contents.find(start_string):\
                        contents.find(end_string)]

text_after_twenty=text_after_twenty.replace('\n',' ')
text_after_twenty=re.sub(r"\s+"," ",text_after_twenty)
text_after_twenty
```

The preceding code generates the following output:

```
'AFTER TWENTY YEARS The policeman on the beat moved up the avenue impressively. The impressiveness was hab
itual and not for show, for spectators were few. The time was barely 10 o\'clock at night, but chilly gust
s of wind with a taste of rain in them had well nigh de-peopled the streets. Trying doors as he went, twir
ling his club with many intricate and artful movements, turning now and then to cast his watchful eye adow
n the pacific thoroughfare, the officer, with his stalwart form and slight swagger, made a fine picture of
a guardian of the peace. The vicinity was one that kept early hours. Now and then you might see the lights
```

Figure 7.6: Text from After Twenty Years

3. Add the following code to extract the required text and print the summarized text, with the **ratio** parameter set to **0.2**:

```
summary_text_after_twenty=summarize(text_after_twenty, \
                                    ratio=0.2)
print(summary_text_after_twenty)
```

The preceding code generates the following output:

```
Now and then you might see the lights of a cigar store or of an all-night lunch counter; but the majority
of the doors belonged to business places that had long since been closed.
About that long ago there used to be a restaurant where this store stands--'Big Joe' Brady's restaurant."
"Until five years ago," said the policeman.
"Twenty years ago to-night," said the man, "I dined here at 'Big Joe' Brady's with Jimmy Wells, my best ch
um, and the finest chap in the world.
```

Figure 7.7: Summarized text when the ratio parameter is 0.2

4. Insert a new cell and add the following code to summarize the text and print the summarized text, with the **ratio** parameter set to **0.25**:

```
summary_text_after_twenty=summarize(text_after_twenty, \
                                    ratio=0.25)
print(summary_text_after_twenty)
```

The preceding code generates the following output:

```
Now and then you might see the lights of a cigar store or of an all-night lunch counter; but the majority
of the doors belonged to business places that had long since been closed.
About that long ago there used to be a restaurant where this store stands--'Big Joe' Brady's restaurant."
"Until five years ago," said the policeman.
"Twenty years ago to-night," said the man, "I dined here at 'Big Joe' Brady's with Jimmy Wells, my best ch
um, and the finest chap in the world.
```

Figure 7.8: Summarized text when the ratio parameter is 0.25

5. Insert a new cell and add the following code to extract the required text from the Wikipedia page for Oscar Wilde:

```
#text_wiki_oscarwilde=wikipedia.summary("Oscar Wilde")
file_url_wiki_oscarwilde=r'../data/oscarwilde/'\
                        'ow_wikipedia_sum.txt'
with open(file_url_wiki_oscarwilde, 'r', \
        encoding='latin-1') as f:
    text_wiki_oscarwilde = f.read()
text_wiki_oscarwilde=text_wiki_oscarwilde.replace('\n',' ')
text_wiki_oscarwilde=re.sub(r"\s+"," ",text_wiki_oscarwilde)
text_wiki_oscarwilde
```

The preceding code generates the following output:

```
'Oscar Fingal O\'Flahertie Wills Wilde (16 October 1854 - 30 November 1900) was an Irish poet and playwrig
ht. After writing in different forms throughout the 1880s, the early 1890s saw him become one of the most
popular playwrights in London. He is best remembered for his epigrams and plays, his novel The Picture of
Dorian Gray, and the circumstances of his criminal conviction for "gross indecency", imprisonment, and ear
ly death at age 46. Wilde\'s parents were successful Anglo-Irish intellectuals in Dublin. A young Wilde le
arned to speak fluent French and German. At university, Wilde read Greats; he demonstrated himself to be a
n exceptional classicist, first at Trinity College Dublin, then at Oxford. He became associated with the e
```

Figure 7.9: Text from the Wikipedia page for Oscar Wilde

6. Insert a new cell and add the following code to summarize the text and print the summarized text using **ratio=0.2**:

```
summary_wiki_oscarwilde=summarize(text_wiki_oscarwilde, \
                            ratio=0.2)
print(summary_wiki_oscarwilde)
```

The preceding code generates the following output:

```
He is best remembered for his epigrams and plays, his novel The Picture of Dorian Gray, and the circumstan
ces of his criminal conviction for "gross indecency", imprisonment, and early death at age 46.
As a spokesman for aestheticism, he tried his hand at various literary activities: he published a book of
poems, lectured in the United States and Canada on the new "English Renaissance in Art" and interior decor
ation, and then returned to London where he worked prolifically as a journalist.
Unperturbed, Wilde produced four society comedies in the early 1890s, which made him one of the most succe
```

Figure 7.10: Summarized text when the ratio parameter is 0.2

7. Add the following code to summarize the text and print the summarized text using **ratio=0.25**:

```
summary_wiki_oscarwilde=summarize(text_wiki_oscarwilde, \
                            ratio=0.25)
print(summary_wiki_oscarwilde)
```

The preceding code generates the following output:

```
He is best remembered for his epigrams and plays, his novel The Picture of Dorian Gray, and the circumstan
ces of his criminal conviction for "gross indecency", imprisonment, and early death at age 46.
As a spokesman for aestheticism, he tried his hand at various literary activities: he published a book of
poems, lectured in the United States and Canada on the new "English Renaissance in Art" and interior decor
ation, and then returned to London where he worked prolifically as a journalist.
Unperturbed, Wilde produced four society comedies in the early 1890s, which made him one of the most succe
```

Figure 7.11: Summarized text when the ratio is 0.25

> **NOTE**
>
> To access the source code for this specific section, please refer
> to https://packt.live/3i5sNQn.
>
> You can also run this example online at https://packt.live/39G0Knx.

We find that the summary for the Wikipedia article is much more coherent than the short story. We can also see that the summary with a **ratio** of **0.20** is a subset of a summary with a **ratio** of **0.25**. Would extractive summarization work better for a children's fairytale than it does for an O. Henry short story? Let's explore this in the next exercise.

EXERCISE 7.03: SUMMARIZING A CHILDREN'S FAIRY TALE USING TEXTRANK

In this exercise, we consider the fairy tale *Little Red Riding Hood* in two variations for the input texts. The first variation is from *Children's Hour with Red Riding Hood and Other Stories*, edited by Watty Piper, while the second variation is from *The Fairy Tales of Charles Perrault*, both of which are available on Project Gutenberg's website. The aim of this exercise is to explore how TextRank (gensim) performs on this summarization.

> **NOTE**
>
> You can find the text from the Watty Piper variation
> at https://packt.live/2Xd30xy. The text from the Charles Perrault
> version can be found at https://packt.live/30g5ZHy.

Complete the following steps to implement this exercise:

1. Open a Jupyter notebook.

2. Insert a new cell and add the following code to import the required libraries:

```
from gensim.summarization import summarize
import re
```

3. Insert a new cell and add the following code to fetch Watty Piper's version of *Little Red Riding Hood*:

```
file_url_grimms=r'../data/littleredrh/pg11592.txt'
with open(file_url_grimms, 'r') as f:
        contents_grimms = f.read()
start_string_grimms='LITTLE RED RIDING HOOD\n\n\n'
end_string_grimms='\n\n\n\n\nTHE GOOSE-GIRL'
text_grimms=contents_grimms[contents_grimms.find(\
                            start_string_grimms):\
                            contents_grimms.find(\
                            end_string_grimms)]
text_grimms=text_grimms.replace('\n',' ')
text_grimms=re.sub(r"\s+"," ",text_grimms)
text_grimms
```

The preceding code generates the following output:

'LITTLE RED RIDING HOOD There was once a sweet little maid who lived with her father and mother in a prett
y little cottage at the edge of the village. At the further end of the wood was another pretty cottage and
in it lived her grandmother. Everybody loved this little girl, her grandmother perhaps loved her most of a
ll and gave her a great many pretty things. Once she gave her a red cloak with a hood which she always wor
e, so people called her Little Red Riding Hood. One morning Little Red Riding Hood\'s mother said, "Put on
your things and go to see your grandmother. She has been ill; take along this basket for her. I have put i
n it eggs, butter and cake, and other dainties." It was a bright and sunny morning. Red Riding Hood was so

Figure 7.12: Text from the Watty Piper variation of Little Red Riding Hood

4. Insert a new cell, add the following code, and fetch the Perrault fairy tale version of *Little Red Riding Hood*:

```
file_url_perrault=r'../data/littleredrh/pg29021.txt'
with open(file_url_perrault, 'r') as f:
        contents_perrault = f.read()
start_string_perrault='Little Red Riding-Hood\n\n'
end_string_perrault='\n\n_The Moral_'
text_perrault=contents_perrault[contents_perrault.find(\
                            start_string_perrault):\
```

```
                                    contents_perrault.find(\
                                    end_string_perrault)]
text_perrault=text_perrault.replace('\n',' ')
text_perrault=re.sub(r"\s+"," ",text_perrault)
text_perrault
```

The preceding code generates the following output:

'Little Red Riding-Hood Once upon a time, there lived in a certain village, a little country girl, the pre
ttiest creature was ever seen. Her mother was excessively fond of her; and her grand-mother doated on her
much more. This good woman got made for her a little red riding-hood; which became the girl so extremely w
ell, that every body called her Little Red Riding-Hood. One day, her mother, having made some girdle-cake
s, said to her: "Go, my dear, and see how thy grand-mamma does, for I hear she has been very ill, carry he
r a girdle-cake, and this little pot of butter." Little Red Riding-Hood set out immediately to go to her g
rand-mother, who lived in another village. As she was going thro\' the wood, she met with Gaffer Wolf, who

Figure 7.13: Tales from the Perrault version of Little Red Riding Hood

5. Insert a new cell and add the following code to generate the two summaries with a **ratio** of **0.20**:

```
llrh_grimms_textrank=summarize(text_grimms,ratio=0.20)
llrh_perrault_textrank=summarize(text_perrault,ratio=0.20)
```

6. Insert a new cell and add the following code to print the TextRank summary (**ratio** of **0.20**) of Grimm's version of *Little Red Riding Hood*:

```
print(llrh_grimms_textrank)
```

The preceding code generates the following output:

LITTLE RED RIDING HOOD There was once a sweet little maid who lived with her father and mother in a pretty
little cottage at the edge of the village.
One morning Little Red Riding Hood's mother said, "Put on your things and go to see your grandmother.
"What have you in that basket, Little Red Riding Hood?" "Eggs and butter and cake, Mr. Wolf." "Where are y
ou going with them, Little Red Riding Hood?" "I am going to my grandmother, who is ill, Mr. Wolf." "Where

Figure 7.14: Output after implementing TextRank on the Watty Piper variation

7. Insert a new cell and add the following code to print the TextRank summary (**ratio** of **0.20**) of Perrault's version of *Little Red Riding Hood*:

```
print(llrh_perrault_textrank)
```

The preceding code generates the following output:

"Who's there?" "Your grand-child, Little Red Riding-Hood," replied the Wolf, counterfeiting her voice, "wh
o has brought you a girdle-cake, and a little pot of butter, sent you by mamma." The good grand-mother, wh
o was in bed, because she found herself somewhat ill, cry'd out: "Pull the peg, and the bolt will fall." T
he Wolf pull'd the peg, and the door opened, and then presently he fell upon the good woman, and ate her u

Figure 7.15: Output after implementing TextRank on the Perrault version

8. Add the following code to generate two summaries with a **ratio** of **0.5**:

```
llrh_grimms_textrank=summarize(text_grimms,ratio=0.5)
llrh_perrault_textrank=summarize(text_perrault,ratio=0.5)
```

9. Add the following code to print a TextRank summary (**ratio** of **0.5**) of Piper's version of *Little Red Riding Hood*:

```
print(llrh_grimms_textrank)
```

The preceding code generates the following output:

```
LITTLE RED RIDING HOOD There was once a sweet little maid who lived with her father and mother in a pretty
little cottage at the edge of the village.
At the further end of the wood was another pretty cottage and in it lived her grandmother.
One morning Little Red Riding Hood's mother said, "Put on your things and go to see your grandmother.
Little Red Riding Hood wandered from her path and was stooping to pick a flower when from behind her a gru
ff voice said, "Good morning, Little Red Riding Hood." Little Red Riding Hood turned around and saw a grea
```

Figure 7.16: Output after implementing TextRank on the Watty Piper variation

10. Add the following code to print a TextRank summary (**ratio** of **0.5**) of Perrault's version of *Little Red Riding Hood*:

```
print(llrh_perrault_textrank)
```

The preceding code generates the following output:

```
One day, her mother, having made some girdle-cakes, said to her: "Go, my dear, and see how thy grand-mamma
does, for I hear she has been very ill, carry her a girdle-cake, and this little pot of butter." Little Re
d Riding-Hood set out immediately to go to her grand-mother, who lived in another village.
The poor child, who did not know that it was dangerous to stay and hear a Wolf talk, said to him: "I am go
ing to see my grand-mamma, and carry her a girdle-cake, and a little pot of butter, from my mamma." "Does
```

Figure 7.17: Output after implementing TextRank on the Perrault version

> **NOTE**
>
> To access the source code for this specific section, please refer to https://packt.live/3i5sRzB.
>
> You can also run this example online at https://packt.live/2XfObu1.

With this, we've found that the four summaries lack coherency and are also incomplete. This is also true of the two summaries with a **ratio** of **0.5**—that is, even when half of the sentences are extracted for the summary. This might be because the conversations in the fairytale are contextual in nature, as a sentence often refers to the preceding sentence(s). This contextual aspect of language makes NLP complex for machines.

Interestingly, extractive summarization works much better for an O. Henry short story such as *After Twenty Years* than it does for a children's fairytale such as *Little Red Riding Hood*. Furthermore, this is not specific to the language used by a specific author, as we have explored with two different variations of this fairytale. It seems a fairytale is unsuitable for extractive summarization. Lets now do an activity in which we'll use the TextRank algorithm to summarize complaints that customers have written against some organizations.

ACTIVITY 7.01: SUMMARIZING COMPLAINTS IN THE CONSUMER FINANCIAL PROTECTION BUREAU DATASET

The Consumer Financial Protection Bureau publishes consumer complaints made against organizations in the financial sector. This original dataset is available at https://www.consumerfinance.gov/data-research/consumer-complaints/#download-the-data. To complete this activity, you will summarize a few complaints using TextRank.

> **NOTE**
>
> You can find the dataset to be used for this activity at
> https://www.dropbox.com/sh/qmq3x3ah1cf3ecz/AAAg_
> E6f0I5vdaB4WVmR6TCga?dl=0&preview=Consumer_Complaints.csv.
> To complete the activity, you will need to place the `.csv` file into the
> `data` folder for this chapter in your local directory.

Follow these steps to implement this activity:

1. Import the summarization libraries and instantiate the summarization model.

2. Load the dataset from a `.csv` file into a pandas DataFrame. Drop all columns other than **Product**, **Sub-product**, **Issue**, **Sub-issue**, and **Consumer complaint narrative**.

3. Select 12 complaints corresponding to the rows **242830**, **1086741**, **536367**, **957355**, **975181**, **483530**, **950006**, **865088**, **681842**, **536367**, **132345**, and **285894** from the 300,000 odd complaints with a narrative. Note that since the dataset is an evolving dataset, the use of a version that's different from the one in the **data** folder could give different results because the input texts could be different.

4. Add a column with the TextRank summary. Each element of this column corresponds to a summary, using TextRank, of the complaint narrative in the corresponding column. Use a **ratio** of **0.20**. Also, use a **try-except** clause since the gensim implementation of the TextRank algorithm throws exceptions with summaries that have very few sentences.

5. Show the DataFrame. You should get an output similar to the following figure:

	Product	Sub-product	Issue	Sub-issue	Consumer complaint narrative	TextRank Summary
132345	Mortgage	FHA mortgage	Loan servicing, payments, escrow account	NaN	I have a trial period of loan modification HAR...	They will not acknowledge my modification, and...
242830	Debt collection	Medical	Cont'd attempts collect debt not owed	Debt is not mine	XXXX XX/XX/XXXX XXXX XXXX, XXXX As a 100 % rat...	XXXX Care coordinator XXXX XXXX contacted the ...
285894	Prepaid card	General purpose card	Managing, opening, or closing account	NaN	I purchased a prepaid credit card from XXXX wh...	Was refused and told I have to use the card.
483530	Debt collection	I do not know	Cont'd attempts collect debt not owed	Debt is not mine	This company has called me XXXX times about a ...	

Figure 7.18: DataFrame showing the summarized complaints

> **NOTE**
>
> The full solution to this activity can be found on page 409.

RECENT DEVELOPMENTS IN TEXT GENERATION AND SUMMARIZATION

Alan Turing (for whom the equivalent of the Nobel Prize in Computer Science is named) proposed a test for artificial intelligence in 1950. This test, known as the Turing Test, says that if humans ask questions and cannot distinguish between text responses generated by a machine and a human, then that machine can be deemed to be intelligent.

Text generation using very large models, such as the GPT-2 (with around 1.5 billion parameters) and **BERT (Bidirectional Encoder Representation from Transformers)** (with around 340 million parameters), can aid in auto-completion tasks. Auto-completion presents unique ethical challenges. While it can offer convenience, it can also reinforce biases in the data. This is accentuated by the fact that most user experience layouts can show only a limited number of options. Furthermore, auto-completion can controversially suggest responses that are different from what the sender originally wants to type.

Unfortunately, most use cases for text generation are negative use cases for generating spam and misinformation. Given that the Turing Test may not be passed any time soon, we are clearly nowhere near considering text generation as a proxy for thought within a machine and there is no widely accepted benchmark for text generation.

Since late 2018, with the invention of self-attention, transformers, and BERT, these approaches are generally considered the best way to teach a machine about some of the most challenging NLP tasks. Self-attention is a technique in which a word is combined with other words in its neighborhood by matrix multiplications. Such multiplications are possible because of vector representations of words. Using such combined representations for all the words in a sentence allows us to represent a sentence in a way that captures context. This allows us to build much larger models that have a significantly higher capacity to learn. A transformer is a combination of attention units and includes position information for each word, that is, multiple self-attention layers and position information are used to capture the context better.

BERT is a transformer that learns the sequential structure of a text in both directions, that is, from left to right and from right to left. This is achieved by randomly masking the words while the model is trained, much like how children are often taught a language by using fill-in-the-blanks exercises. Such is the generalized learning of BERT that it can be used even for translation-related tasks, even though it has not been specifically taught translation as a task. BERT and other large models, such as GPT-2, require a huge computing infrastructure, which is generally not available to most people outside of leading universities and the biggest technology corporations. Pre-trained models fill the void in such cases. The TextRank algorithm considers each sentence to be a bag of words. With the advent of BERT, it is possible for us to have a superior sentence representation that captures meaning much better than the bag of words model.

In the case of summarization, even though there is a benchmark called **Recall-Oriented Understudy for Gisting Evaluation** (**ROUGE**), summarization is best evaluated qualitatively given that there isn't only one correct way to summarize text. In February 2020, Microsoft's Turing NLG model, which has 17 billion parameters, generated abstractive summaries for three examples, which were shared publicly. However, the model is not publicly available currently and so the results cannot be reproduced.

Furthermore, we don't know how the Microsoft NLG model does with a naïve test such as the Little Red Riding Hood test. In general, extractive summarization of the kind discussed earlier in this chapter is by far the most useful for practitioners compared with the utility of the state-of-the-art technology in text generation and paraphrasing. Due to this, in the next section, we'll largely focus on practical challenges in extractive summarization.

PRACTICAL CHALLENGES IN EXTRACTIVE SUMMARIZATION

Given the rapid pace of development in NLP, it is even more important to use compatible versions of the libraries that we use. Evaluation of a document's suitability for extractive summarization can be undertaken manually. Often, we would like to summarize multiple pieces of text, all of which could be short in length. The TextRank algorithm will not work well in such cases.

All unverified claims reported in this field ought to be taken with a grain of salt until the claim has been verified. Such claims ought to be subjected by practitioners to naïve tests such as the Little Red Riding test. We can only use a model if it works and if the limitations related to scope and any biases are considered.

SUMMARY

In this chapter, we learned about text generation using Markov chains and extractive summarization using the TextRank algorithm. We also explored both the power and limitations of various advanced approaches. In the next chapter, we will learn about sentiment analysis.

8

SENTIMENT ANALYSIS

OVERVIEW

This chapter introduces you to one of the most exciting applications of natural language processing—that is, sentiment analysis. You will explore the various tools used to perform sentiment analysis, such as popular NLP libraries and deep learning frameworks. You will then perform sentiment analysis on given text data using the powerful `textblob` library. You will load textual data and perform preprocessing on it to fine-tune the results of your sentiment analysis program. By the end of the chapter, you will be able to train a sentiment analysis model.

INTRODUCTION

In the previous chapter, we looked at text generation, paraphrasing, and summarization, all of which can be immensely useful in helping us focus on only the essential and meaningful parts of the text corpus. This, in turn, helps us to further refine the results of our NLP project. In this chapter, we will look at **sentiment analysis**, which, as the name suggests, is the area of NLP that involves teaching computers how to identify the sentiment behind written content or parsed audio—that is, audio converted to text. Adding this ability to automatically detect sentiment in large volumes of text and speech opens new possibilities for us to write useful software.

In sentiment analysis, we try to build models that detect how people feel. This starts with determining what kind of feeling we want to detect. Our application may attempt to determine the level of human emotion (most often, whether a person is sad or happy; satisfied or dissatisfied; or interested or disinterested and so on). The common thread here is that we measure how sentiments vary in different directions. This is also called polarity. Polarity signifies the emotions present in a sentence, such as joy or anger. For example, "I love oranges" implies an emotionally positive statement, whereas "I hate politics" is a strong negative emotion.

WHY IS SENTIMENT ANALYSIS REQUIRED?

In machine learning projects, we try to build applications that work similarly to a human being. We measure success in part by seeing how close our application is to matching human-level performance. Generally, machine learning programs cannot exceed human-level performance by a significant margin—especially if our training data source is human-generated.

Let's say that we want to carry out a sentiment analysis of product reviews. The sentiment analysis program should detect how reviewers feel. Obviously, it is impractical for a person to read thousands of movie reviews. This is where automated sentiment analysis enters the picture. Artificial intelligence is useful when it is impractical for people to perform the task. In this case, the task is reading thousands of reviews.

THE GROWTH OF SENTIMENT ANALYSIS

The field of sentiment analysis is driven by a few main factors. Firstly, it's driven by the rapid growth in online content that's used by companies to understand and respond to how people feel. Secondly, since sentiment drives human decisions, businesses that understand their customers' sentiments have a major advantage in predicting and shaping purchasing decisions. Finally, NLP technology has improved significantly, allowing the much wider application of sentiment analysis.

THE MONETIZATION OF EMOTION

The growth of the internet and internet services has enabled new business models to work with human connection, communication, and sentiment. In January 2020, Facebook had about 61.3% of the social media traffic and has been one of the most successful social media platforms at connecting people across the world and providing features that enable users to express their thoughts and post memorable moments from their life online. Similarly, although Twitter had just 14.51% of the traffic, it has still proved to be an influential way to display sentiment online.

There are now large amounts of information on social media about what people like or dislike. This data is of significant value not only in business but also in political campaigns. This means that sentiment has significant business value and can be monetized.

TYPES OF SENTIMENTS

There are various sentiments that we can try to detect in language sources. Let's discuss a few of them in detail.

EMOTION

Sentiment analysis is often used to detect the emotional state of a person. It checks whether the person is happy or sad, or content or discontent. Businesses often use it to improve customer satisfaction. For example, let's look at the following statement:

"I thought I would have enjoyed the movie, but it left me feeling that it could have been better."

In this statement, it seems as though the person who has just watched a movie is unhappy about it. A sentiment detector, in this case, would be able to classify the review as negative and allow the business (the movie studio, for instance) to adjust how they make movies in the future.

Action Orientation versus Passivity

This is about whether a person is prone to action or not. This is often used to determine how close a person is to making a choice. For example, using a travel reservation chatbot, you can detect whether a person needs to make a reservation urgently or is simply making passive queries and is therefore less likely to book a ticket right now. The level of action orientation or passivity provides additional clues to detect intention. This can be used to make smart business decisions.

Tone

Speech and text are often meant to convey certain impressions that are not necessarily factual and not entirely emotional. Examples of this are sarcasm, irony, and humor. This may provide useful additional information about how a person thinks. Although tone is tricky to detect, there might be certain words or phrases that are often used in certain contexts. We can use NLP algorithms to extract statistical patterns from document sources. For example, we can use sentiment analysis to detect whether a news article is sarcastic.

Subjectivity versus Objectivity

You may want to detect whether the given text source is subjective or objective. For example, you might want to detect whether a person has issued and expressed an opinion, or whether their statement reads more like a fact and can only be true or false. Let's look at the following two statements to get a better understanding:

- Statement 1: "The duck was overcooked, and I could hardly taste the flavor."

- Statement 2: "Ducks are aquatic birds."

In these two statements, statement 1 should be recognized as a subjective opinion and statement 2 as an objective fact. Determining the objectivity of a statement helps us decide on the appropriate response to the statement.

KEY IDEAS AND TERMS

Let's look at some of the key ideas and terms that are used in sentiment analysis.

Classification

As we learned in *Chapter 3, Developing a Text Classifier*, classification is the NLP technique of assigning one or more classes to text documents. This helps in separating and sorting the documents. If you use classification for sentiment analysis, you assign different sentiment classes such as positive, negative, or neutral. Sentiment analysis is a type of text classification that aims to create a classifier trained on a set of labeled pairs – text and its corresponding sentiment (label). Upon training such a classifier on a large labeled dataset, the sentiment analysis model generalizes well and can classify unseen text into appropriate sentiment categories.

Supervised Learning

As we have already seen, in **supervised learning**, we create a model by supplying data and labeled targets to the training algorithms. The algorithms learn using this supply. When it comes to sentiment analysis, we provide the training dataset with the labels that represent the sentiment. For example, for each text in a dataset, we would assign a value of 1 if the sentiment is positive, and a value of 0 if the statement is negative.

Polarity

Polarity is a measure of how negative or positive the sentiment is in a given language. Polarity is used because it is simple and easy to measure and can be easily translated to a simple numeric scale. It usually ranges between -1 and 1. Values close to 1 reflect documents that have positive sentiments, whereas values close to -1 reflect documents that have negative sentiments. Values around 0 reflect documents that are neutral in sentiment.

It's worth noting that the polarity detected by a model depends on how it has been trained. On political Reddit threads, the opinions tend to be highly polarized. On the other hand, if you use the same model on business documents to measure sentiments, the scores tend to be neutral. So, you need to choose models that are trained in similar domains.

Intensity

In contrast to polarity, which is measured from negative to positive, intensity is measured in terms of arousal, which ranges from low to high. Most often, the level of intensity is included in the sentiment score. It is measured by looking at the closeness of the score to 0 or 1.

APPLICATIONS OF SENTIMENT ANALYSIS

There are various applications of sentiment analysis.

Financial Market Sentiment

Financial markets operate partially on economic fundamentals but are also heavily influenced by human sentiment. Stock market prices, which tend to rise and fall, are influenced by the opinions of news articles regarding the overall market or any specific securities.

Financial market sentiment helps measure the overall attitude of investors toward securities. Market sentiment can be detected using news or social media articles. We can use NLP algorithms to build models that detect market sentiment and use those models to predict future market prices.

Product Satisfaction

Sentiment analysis is commonly used to determine how customers feel about products and services. For example, Amazon makes use of its extensive product reviews dataset. This not only helps to improve its products and services but also acts as a source of training data for its sentiment analysis services.

Social Media Sentiment

A really useful area of focus for sentiment analysis is social media monitoring. Social media has become a key communication medium with which most people around the world interact every day, and so there is a large and growing source of human language data available there. More importantly, the need for businesses and organizations to be able to process and understand what people are saying on social media has only increased. This has led to an exponential growth in demand for sentiment analysis services.

Brand Monitoring

A company's brand is a significant asset and companies spend a lot of time, effort, and money maintaining their brand value. With the growth of social media, companies are now exposed to considerable potential brand risks from negative social media conversations. On the other hand, there is also the potential for positive brand growth from positive interactions and messages on social media. For this reason, businesses deploy people to monitor what is said about them and their brands on social media. Automated sentiment analysis makes this significantly easier and also more efficient.

Customer Interaction

Organizations often want to know how their customers feel during an interaction in an online chat or a phone conversation. In such cases, the objective is to detect the level of satisfaction with the service or the products. Sentiment analysis tools help companies handle large volumes of text and voice data that are generated during customer interaction. Every company, irrespective of the domain, wants to utilize the data at their disposal to glean valuable insights, as there is potential revenue to be had if companies can gain insights into customer satisfaction.

TOOLS USED FOR SENTIMENT ANALYSIS

There are a lot of tools capable of analyzing sentiment. Each tool has its advantages and disadvantages. We will look at each of them in detail.

NLP SERVICES FROM MAJOR CLOUD PROVIDERS

Online sentiment analysis is carried out by all major cloud services providers, such as Amazon, Microsoft, Google, and IBM. You can usually find sentiment analysis as a part of their text analysis services or general machine learning services. Online services offer the convenience of packaging all the necessary algorithms behind the provider's API. These algorithms are capable of performing sentiment analysis. To use such services, you need to provide the text or audio sources, and in return, the services will provide you with a measure of the sentiment. These services usually return a standard, simple score, such as positive, negative, or neutral. The score usually ranges between 0 and 1.

The following are the advantages and disadvantages of NLP services from major cloud providers:

Advantages

- You require almost no knowledge of NLP algorithms or sentiment analysis. This results in fewer staffing needs.

- Sentiment analysis services provide their own computation, reducing your own computational infrastructure needs.

- Online services can scale well beyond what regular companies can do on their own.

- You gain the benefits of automatic improvements and updates to sentiment analysis algorithms and data.

Disadvantages

- Online services require—at least temporarily—a reduction in privacy since you must provide the documents to be analyzed by the service. Depending on your project's privacy needs, this may or may not be acceptable. There might also be laws that restrict data crossing into another national jurisdiction.

- The service provided by cloud providers is like one-solution-fits-all and is considered very generic, so it won't necessarily apply to niche use cases.

ONLINE MARKETPLACES

Recently, AI marketplaces have emerged that offer different algorithms from third parties. Online marketplaces differ from cloud providers. An online marketplace allows third-party developers to deploy sentiment analysis services on their platform.

Here are the advantages and disadvantages of online marketplaces:

Advantages

- AI marketplaces provide the flexibility of choosing between different sentiment analysis algorithms instead of just one algorithm. This enables users to try out different techniques and see which one fits their business needs the best.

- Using algorithms from an AI marketplace reduces the need for dedicated data scientists for your project.

Disadvantages

- Algorithms from third parties are of varying quality.

- Since the algorithms are provided by smaller companies, there is no guarantee that they won't disappear. And for businesses, this is a big risk since their solution has a direct dependency on a third party that is outside their control.

PYTHON NLP LIBRARIES

There are a few NLP libraries that need to be integrated into your project instead of being called upon as services. These are called dedicated NLP libraries and they usually include many NLP algorithms from academic research. Sophisticated NLP libraries used across the industry are spaCy, gensim, and AllenNLP.

Here are the advantages and disadvantages of Python NLP libraries:

Advantages

- It's usually state-of-the-art research that goes into these libraries, and they usually have well-chosen datasets.

- They provide a framework that makes it much easier to build projects and do rapid experiments.

- They offer out-of-the-box abstractions that are required for all NLP projects, such as Token and Span.

- They are easy to scale to real-world deployment.

Disadvantages

- This won't be considered a true disadvantage since libraries are meant to be general-purpose, but for complex use cases, developers would have to write their own implementations as required.

DEEP LEARNING FRAMEWORKS

Deep learning libraries such as PyTorch and TensorFlow are meant to be used to build complex models for a wide range of applications, not limited to just NLP. These libraries provide you with more advanced algorithms and mathematical functions, helping you develop powerful and complex models.

The advantages and disadvantages of these frameworks are explained here:

Advantages

- You have the flexibility to develop your sentiment analysis model to meet complex business needs.

- You can integrate the latest and the most advanced algorithms when they are available in general-purpose libraries.

- You can make use of transfer learning, which takes a model trained on a large text source, to fine-tune the training as per your project's needs. This allows you to create a sentiment analysis model that is more suitable for your needs.

Disadvantages

- This approach requires you to have in-depth knowledge of machine learning and complex topics such as deep learning.

- Deep learning libraries require a large volume of rich annotated datasets along with an intense computational infrastructure to train and experiment with different modeling techniques to get a generalized model that's fit to be deployed in production. So, there is a requirement for training on non-CPU hardware such as GPUs/TPUs.

Now that we've learned about the various tools available for sentiment analysis, let's explore the most popular Python libraries.

THE TEXTBLOB LIBRARY

textblob is a Python library used for NLP, as we've seen in the previous chapters. It has a simple API and is probably the easiest way to begin with sentiment analysis. **textblob** is built on top of the NLTK library but is much easier to use. In the following sections, we will do an exercise and an activity to get a better understanding of how we can use **textblob** for sentiment analysis.

EXERCISE 8.01: BASIC SENTIMENT ANALYSIS USING THE TEXTBLOB LIBRARY

In this exercise, we will perform sentiment analysis on a given text. For this, we will be using the **TextBlob** class of the **textblob** library. Follow these steps to complete this exercise:

1. Open a Jupyter notebook.

2. Insert a new cell and add the following code to implement to import the **TextBlob** class from the **textblob** library:

```
from textblob import TextBlob
```

3. Create a variable named **sentence** and assign it a string. Insert a new cell and add the following code to implement this:

```
sentence = "but you are Late Flight again!! "\
           "Again and again! Where are the  crew?"
```

4. Create an object of the **TextBlob** class. Add **sentence** as a parameter to the **TextBlob** container. Insert a new cell and add the following code to implement this:

```
blob = TextBlob(sentence)
```

5. In order to view the details of the **blob** object, insert a new cell and add the following code:

```
print(blob)
```

The code generates the following output:

```
but you are Late Flight again!! Again and again! Where are the crew?
```

6. To use the **sentiment** property of the **TextBlob** class (which returns a tuple), insert a new cell and add the following code:

```
blob.sentiment
```

The code generates the following output:

```
Sentiment(polarity=-0.5859375, subjectivity=0.6
```

> **NOTE**
>
> To access the source code for this specific section, please refer to https://packt.live/2DlQvbM.
>
> You can also run this example online at https://packt.live/3jXvAN1.

In the code, we can see the **polarity** and **subjectivity** scores for a given text. The output indicates a polarity score of -0.5859375, which means that negative sentiment has been detected in the text. The subjectivity score means that the text is somewhat on the subjective side, though not entirely subjective. We have performed sentiment analysis on a given text using the **textblob** library. In the next section, we will perform sentiment analysis on tweets about airlines.

ACTIVITY 8.01: TWEET SENTIMENT ANALYSIS USING THE TEXTBLOB LIBRARY

In this activity, you will perform sentiment analysis on tweets related to airlines. You will also be providing condition for determining positive, negative, and neutral tweets, using the **textblob** library.

> **NOTE**
>
> You can find the data to be used for this activity here: https://packt.live/33cnr1q.

Follow these steps to implement this activity:

1. Import the necessary libraries.

2. Load the CSV file.

3. Fetch the **text** column from the DataFrame.

4. Extract and remove the handles from the fetched data.

5. Perform sentiment analysis and get the new DataFrame.

6. Join both the DataFrames.

7. Apply the appropriate conditions and view positive, negative, and neutral tweets.

After executing those steps, the output for positive tweets should be as follows:

	tweet	At	tweet_preprocessed	Polarity	Subjectivity
8	@FlyHighAirways Well, I didn't...but NOW I DO! :-D	@FlyHighAirways	Well, I didn't...but NOW I DO! :-D	1.000000	1.000000
19	@FlyHighAirways you know what would be amazingly awesome? BOS-FLL PLEASE!!!!!!! I want to fly with only you.	@FlyHighAirways	you know what would be amazingly awesome? BOS-FLL PLEASE!!!!!!! I want to fly with only you.	0.600000	0.966667
22	@FlyHighAirways I love the hipster innovation. You are a feel good brand.	@FlyHighAirways	I love the hipster innovation. You are a feel good brand.	0.600000	0.600000

Figure 8.1: Positive tweets

As you can see from the preceding output, the **Polarity** column shows a positive integer. This implies that the tweet displays positive sentiment. The **Subjectivity** column indicates that most tweets are found to be of a subjective nature.

The output for negative tweets is as follows:

	tweet	At	tweet_preprocessed	Polarity	Subjectivity
33	@FlyHighAirways awaiting my return phone call, just would prefer to use your online self-service option :(	@FlyHighAirways	awaiting my return phone call, just would prefer to use your online self-service option :(	-0.750000	1.000000
84	@FlyHighAirways it was a disappointing experience which will be shared with every business traveler I meet. #neverflyFlyHigh	@FlyHighAirways	it was a disappointing experience which will be shared with every business traveler I meet. #neverflyFlyHigh	-0.600000	0.700000
114	@FlyHighAirways come back to #PHL already. We need you to take us out of this horrible cold. #pleasecomeback http://t.co/gLXFwP6nQH	@FlyHighAirways	come back to #PHL already. We need you to take us out of this horrible cold. #pleasecomeback http://t.co/gLXFwP6nQH	-0.533333	0.666667

Figure 8.2: Negative tweets

The preceding output shows a **`Polarity`** column with a negative integer, implying that the tweet displays negative sentiment, while the **`Subjectivity`** column shows a positive integer, which implies the same as before—personal opinion or feeling.

The output for neutral tweets should be as follows:

	tweet	At	tweet_preprocessed	Polarity	Subjectivity
0	@FlyHighAirways What @dhepburn said.	@FlyHighAirways	What said.	0.00000	0.00
1	@FlyHighAirways plus you've added commercials to the experience... tacky.	@FlyHighAirways	plus you've added commercials to the experience... tacky.	0.00000	0.00
3	@FlyHighAirways it's really aggressive to blast obnoxious "entertainment" in your guests' faces & they have little recourse	@FlyHighAirways	it's really aggressive to blast obnoxious "entertainment" in your guests' faces & they have little recourse	0.00625	0.35

Figure 8.3: Neutral tweets

The preceding output has a **Polarity** column and a **Subjectivity** column with a zero or almost zero value. This implies the tweet has neither positive nor negative sentiment, but neutral; moreover, no subjectivity is detected for these tweets.

> **NOTE**
>
> The solution to this activity can be found on page 412.

In the next section, we will explore more about performing sentiment analysis using online web services.

UNDERSTANDING DATA FOR SENTIMENT ANALYSIS

Sentiment analysis is a type of **text classification**. Sentiment analysis models are usually trained using **supervised datasets**. Supervised datasets are a kind of dataset that is labeled with the target variable, usually a column that specifies the sentiment value in the text. This is the value we want to predict in the unseen text.

EXERCISE 8.02: LOADING DATA FOR SENTIMENT ANALYSIS

In this exercise, we will load data that could be used to train a sentiment analysis model. For this exercise, we will be using three datasets—namely Amazon, Yelp, and IMDb.

> **NOTE**
>
> You can find the data being used in this exercise here:
> https://packt.live/2XgeQql.

Follow these steps to implement this exercise:

1. Open a Jupyter notebook.

2. Insert a new cell and add the following code to import the necessary libraries:

```
import pandas as pd
pd.set_option('display.max_colwidth', 200)
```

This imports the **pandas** library. It also sets the display width to **200** characters so that more of the review text is displayed on the screen.

3. To specify where the sentiment data is located, first load three different datasets from Yelp, IMDb, and Amazon. Insert a new cell and add the following code to implement this:

```
DATA_DIR = 'data/sentiment_labelled_sentences/'
IMDB_DATA_FILE = DATA_DIR + 'imdb_labelled.txt'
YELP_DATA_FILE = DATA_DIR + 'yelp_labelled.txt'
AMAZON_DATA_FILE = DATA_DIR + 'amazon_cells_labelled.txt'
COLUMN_NAMES = ['Review', 'Sentiment']
```

Each of the data files has two columns: one for the review text and a numeric column for the sentiment.

4. To load the IMDb reviews, insert a new cell and add the following code:

```
imdb_reviews = pd.read_table(IMDB_DATA_FILE, names=COLUMN_NAMES)
```

In this code, the **read_table()** method loads the file into a DataFrame.

5. Display the top **10** records in the DataFrame. Add the following code in the new cell:

```
imdb_reviews.head(10)
```

The code generates the following output:

	Review	Sentiment
0	A very, very, very slow-moving, aimless movie about a distressed, drifting young man.	0
1	Not sure who was more lost - the flat characters or the audience, nearly half of whom walked out.	0
2	Attempting artiness with black & white and clever camera angles, the movie disappointed - became even more ridiculous - as the acting was poor and the plot and lines almost non-existent.	0
3	Very little music or anything to speak of.	0
4	The best scene in the movie was when Gerardo is trying to find a song that keeps running through his head.	1
5	The rest of the movie lacks art, charm, meaning... If it's about emptiness, it works I guess because it's empty.	0
6	Wasted two hours.	0
7	Saw the movie today and thought it was a good effort, good messages for kids.	1
8	A bit predictable.	0
9	Loved the casting of Jimmy Buffet as the science teacher.	1

Figure 8.4: The first few records in the IMDb movie review file

In the preceding figure, you can see that the negative reviews have sentiment scores of **0** and positive reviews have sentiment scores of **1**.

6. To check the total number of records of the IMDb review file, use the **value_counts()** function. Add the following code in a new cell to implement this:

```
imdb_reviews.Sentiment.value_counts()
```

The expected output with total reviews should be as follows:

```
1          386
0          362
Name:    Sentiment, dtype: int64
```

In the preceding figure, you can see that the data file contains a total of **748** reviews, out of which **362** are negative and **386** are positive.

7. Format the data by adding the following code in a new cell:

```
imdb_counts = imdb_reviews.Sentiment.value_counts().to_frame()
imdb_counts.index = pd.Series(['Positive', 'Negative'])
imdb_counts
```

The code generates the following output:

	Sentiment
Positive	386
Negative	362

Figure 8.5: Counts of positive and negative sentiments in the IMDb review file

We called **value_counts()**, created a DataFrame, and assigned **Positive** and **Negative** as index labels.

8. To load the Amazon reviews, insert a new cell and add the following code:

```
amazon_reviews = pd.read_table(AMAZON_DATA_FILE, \
                            names=COLUMN_NAMES)
amazon_reviews.head(10)
```

The code generates the following output:

	Review	Sentiment
0	So there is no way for me to plug it in here in the US unless I go by a converter.	0
1	Good case, Excellent value.	1
2	Great for the jawbone.	1
3	Tied to charger for conversations lasting more than 45 minutes.MAJOR PROBLEMS!!	0
4	The mic is great.	1
5	I have to jiggle the plug to get it to line up right to get decent volume.	0
6	If you have several dozen or several hundred contacts, then imagine the fun of sending each of them one by one.	0
7	If you are Razr owner...you must have this!	1
8	Needless to say, I wasted my money.	0
9	What a waste of money and time!.	0

Figure 8.6: Reviews from the Amazon dataset

9. To load the Yelp reviews, insert a new cell and add the following code:

```
yelp_reviews = pd.read_table(YELP_DATA_FILE, \
                          names=COLUMN_NAMES)
yelp_reviews.head(10)
```

The code generates the following output:

	Review	Sentiment
0	Wow... Loved this place.	1
1	Crust is not good.	0
2	Not tasty and the texture was just nasty.	0
3	Stopped by during the late May bank holiday off Rick Steve recommendation and loved it.	1
4	The selection on the menu was great and so were the prices.	1
5	Now I am getting angry and I want my damn pho.	0
6	Honeslty it didn't taste THAT fresh.)	0
7	The potatoes were like rubber and you could tell they had been made up ahead of time being kept under a warmer.	0
8	The fries were great too.	1
9	A great touch.	1

Figure 8.7: Reviews from the Yelp dataset

> **NOTE**
>
> To access the source code for this specific section, please refer to https://packt.live/2XfwmLB.
>
> You can also run this example online at https://packt.live/339NTss.

We have learned how to load data that could be used to train a sentiment analysis model. The review files mentioned in this exercise are an example. Each file contains review text, plus a sentiment label for each. This is the minimum requirement of a supervised machine learning project: to build a model that is capable of predicting sentiments. However, the review text cannot be used as is; it needs to be preprocessed so that we can extract feature vectors out of it and eventually provide it as input to the model.

Now that we have learned about loading the data, in the next section, we will focus on training sentiment models.

TRAINING SENTIMENT MODELS

The end product of any sentiment analysis project is a **sentiment model**. This is an object containing a stored representation of the data on which it was trained. Such a model has the ability to predict sentiment values for text that it has not seen before. To develop a sentiment analysis model, the following steps should be taken:

1. The document dataset must be split into train and test datasets. The test dataset is normally a fraction of the overall dataset. It is usually between 5% and 40% of the overall dataset, depending on the total number of examples available. If the amount of data is too large, then a smaller test dataset can be used.

2. Next, the text should be preprocessed by stripping unwanted characters, removing stop words, and performing other common preprocessing steps.

3. The text should be converted to numeric vector representations in order to extract the features. These representations are used for training machine learning models.

4. Once we have the vector representations, we can train the model. This will be specific to the type of algorithm being used. During the training, our model will use the test dataset as a guide to learn about the text.

5. We can then use the model to predict the sentiment of documents that it has not seen before. This is the step that will be performed in production.

In the next section, we will train a sentiment model. We'll make use of the **TfidfVectorizer** and **LogisticRegression** classes, which we explored in one of the previous chapters.

ACTIVITY 8.02: TRAINING A SENTIMENT MODEL USING TFIDF AND LOGISTIC REGRESSION

To complete this activity, you will build a sentiment analysis model using the Amazon, Yelp, and IMDb datasets that you used in the previous exercise. Use the TFIDF method to extract features from the text and use logistic regression for the learning algorithm. The following steps will help you complete this activity:

1. Open a Jupyter notebook.

2. Import the necessary libraries.

3. Load the Amazon, Yelp, and IMDb datasets.

4. Concatenate the datasets and take out a random sample of 10 items.

5. Create a function for preprocessing the text, that is, convert the words into lowercase and normalize them.

6. Apply the function created in the previous step on the dataset.

7. Use **TfidfVectorizer** to convert the review text into TFIDF vectors and use the **LogisticRegression** class to create a model that uses logistic regression for the model. These should be combined into a **Pipeline** object.

8. Now split the data into train and test sets, using 70% to train the data and 30% to test the data.

9. Use the **fit()** function to fit the training data on the pipeline.

10. Print the accuracy score.

11. Test the model on these sentences: *"I loved this place"* and *"I hated this place"*.

> **NOTE**
>
> The full solution to this activity can be found on page 418.

SUMMARY

We started our journey into NLP with basic text analytics and text preprocessing techniques, such as tokenization, stemming, lemmatization, and lowercase conversion, to name a few. We then explored ways in which we can represent our text data in numerical form so that it can be understood by machines in order to implement various algorithms. After getting some practical knowledge of topic modeling, we moved on to text vectorization, and finally, in this chapter, we explored various applications of sentiment analysis. This included different tools that use sentiment analysis, from technologies available from online marketplaces to deep learning frameworks. More importantly, we learned how to load data and train our model to use it to predict sentiment.

APPENDIX

CHAPTER 1: INTRODUCTION TO NATURAL LANGUAGE PROCESSING

ACTIVITY 1.01: PREPROCESSING OF RAW TEXT

Solution

Let's perform preprocessing on a text corpus. To complete this activity, follow these steps:

1. Open a Jupyter Notebook.

2. Insert a new cell and add the following code to import the necessary libraries:

```
from nltk import download
download('stopwords')
download('wordnet')
nltk.download('punkt')
download('averaged_perceptron_tagger')
from nltk import word_tokenize
from nltk.stem.wordnet import WordNetLemmatizer
from nltk.corpus import stopwords
from autocorrect import Speller
from nltk.wsd import lesk
from nltk.tokenize import sent_tokenize
from nltk import stem, pos_tag
import string
```

3. Read the content of **file.txt** and store it in a variable named **sentence**. Insert a new cell and add the following code to implement this:

```
#load the text file into variable called sentence
sentence = open("../data/file.txt", 'r').read()
```

4. Apply tokenization on the given text corpus. Insert a new cell and add the following code to implement this:

```
words = word_tokenize(sentence)
```

5. To print the list of tokens, insert a new cell and add the following code:

```
print(words[0:20])
```

This code generates the following output:

```
['The', 'reader', 'of', 'this', 'course', 'should', 'have',
 'a', 'basic', 'knowledge', 'of', 'the', 'Python', 'programming',
 'lenguage', '.', 'He/she', 'must', 'have', 'knowldge']
```

In the preceding figure, we can see the initial 20 tokens of our text corpus.

6. To perform spelling correction in our given text corpus, loop through each token and correct the tokens that are wrongly spelled. Insert a new cell and add the following code to implement this:

```
spell = Speller(lang='en')

def correct_sentence(words):
    corrected_sentence = ""
    corrected_word_list = []
    for wd in words:
        if wd not in string.punctuation:
            wd_c = spell(wd)
            if wd_c != wd:
                print(wd+" has been corrected to: "+wd_c)
                corrected_sentence = corrected_sentence+" "+wd_c
                corrected_word_list.append(wd_c)
            else:
                corrected_sentence = corrected_sentence+" "+wd
                corrected_word_list.append(wd)
        else:
            corrected_sentence = corrected_sentence + wd
            corrected_word_list.append(wd)
    return corrected_sentence, corrected_word_list

corrected_sentence, corrected_word_list = correct_sentence(words)
```

This code generates the following output:

```
lenguage has been corrected to: language
knowldge has been corrected to: knowledge
Familiarity has been corrected: familiarity
```

7. To print the corrected text corpus, add a new cell and type the following code:

```
corrected_sentence
```

This code generates the following output:

```
' The reader of this course should have a basic knowledge of the
Python programming language. He/she must have knowledge of data
types in Python. He should be able to write functions, and also have
the ability to import and use libraries and packages in Python.
familiarity with basic linguistics and probability is assumed
although not required to fully complete this course.'
```

8. To print a list of the initial 20 tokens of the corrected words, insert a new cell and add the following code:

```
print(corrected_word_list[0:20])
```

This code generates the following output:

```
['The', 'reader', 'of', 'this', 'course', 'should', 'have',
 'a', 'basic', 'knowledge', 'of', 'the', 'Python', 'programming',
 'language', '. ', 'He/she', 'must', 'have', 'knowledge']
```

9. To add a PoS tag to all the corrected words in the list, insert a new cell and add the following code:

```
print(pos_tag(corrected_word_list))
```

This code generates the following output:

```
[('The', 'DT'), ('reader', 'NN'), ('of', 'IN'), ('this', 'DT'), ('course', 'NN'), ('should', 'MD'), ('have', 'V
B'), ('a', 'DT'), ('basic', 'JJ'), ('knowledge', 'NN'), ('of', 'IN'), ('the', 'DT'), ('Python', 'NNP'), ('programm
ing', 'NN'), ('language', 'NN'), ('.', '.'), ('He/she', 'NNP'), ('must', 'MD'), ('have', 'VB'), ('knowledge', 'N
N'), ('of', 'IN'), ('data', 'NNS'), ('types', 'NNS'), ('in', 'IN'), ('Python', 'NNP'), ('.', '.'), ('He', 'PRP'),
('should', 'MD'), ('be', 'VB'), ('able', 'JJ'), ('to', 'TO'), ('write', 'VB'), ('functions', 'NNS'), (',', ','),
('and', 'CC'), ('also', 'RB'), ('have', 'VBP'), ('the', 'DT'), ('ability', 'NN'), ('to', 'TO'), ('import', 'NN'),
('and', 'CC'), ('use', 'NN'), ('libraries', 'NNS'), ('and', 'CC'), ('packages', 'NNS'), ('in', 'IN'), ('Python',
'NNP'), ('.', '.'), ('familiarity', 'NN'), ('with', 'IN'), ('basic', 'JJ'), ('linguistics', 'NNS'), ('and', 'CC'),
('probability', 'NN'), ('is', 'VBZ'), ('assumed', 'VBN'), ('although', 'IN'), ('not', 'RB'), ('requered', 'VBN'),
('to', 'TO'), ('fully', 'RB'), ('complete', 'VB'), ('this', 'DT'), ('course', 'NN'), ('.', '.')]
```

Figure 1.5: List of corrected words tagged with appropriate PoS

10. To remove the stop words, insert a new cell and add the following code:

```
stop_words = stopwords.words('english')
def remove_stop_words(word_list):
    corrected_word_list_without_stopwords = []
    for wd in word_list:
        if wd not in stop_words:
            corrected_word_list_without_stopwords.append(wd)
    return corrected_word_list_without_stopwords

corrected_word_list_without_stopwords = remove_stop_words\
                              (corrected_word_list)
corrected_word_list_without_stopwords[:20]
```

This code generates the following output:

```
['The',
 'reader',
 'course',
 'basic',
 'knowledge',
 'Python',
 'programming',
 'language',
 '.',
 'He/she',
 'must',
 'knowledge',
 'data',
 'types',
 'Python',
 '.',
 'He',
 'able',
 'write',
 'functions']
```

Figure 1.6: List excluding the stop words

In the preceding figure, we can see that the stop words have been removed and a new list has been returned.

11. Apply the stemming process, and then insert a new cell and add the following code:

```
stemmer = stem.PorterStemmer()
def get_stems(word_list):
    corrected_word_list_without_stopwords_stemmed = []
    for wd in word_list:
        corrected_word_list_without_stopwords_stemmed\
        .append(stemmer.stem(wd))
    return corrected_word_list_without_stopwords_stemmed

corrected_word_list_without_stopwords_stemmed = \
get_stems(corrected_word_list_without_stopwords)
corrected_word_list_without_stopwords_stemmed[:20]
```

This code generates the following output:

```
['the',
 'reader',
 'cours',
 'basic',
 'knowledg',
 'python',
 'program',
 'languag',
 '.',
 'he/sh',
 'must',
 'knowledg',
 'data',
 'type',
 'python',
 '.',
 'He',
 'abl',
 'write',
 'function']
```

Figure 1.7: List of stemmed words

In the preceding code, we looped through each of the words in the **corrected_word_list_without_stopwords** list and applied stemming to them. The preceding figure shows the list of the initial 20 stemmed words.

12. To apply the lemmatization process to the corrected word list, insert a new cell and add the following code:

```
lemmatizer = WordNetLemmatizer()
def get_lemma(word_list):
    corrected_word_list_without_stopwords_lemmatized = []
    for wd in word_list:
        corrected_word_list_without_stopwords_lemmatized\
          .append(lemmatizer.lemmatize(wd))
    return corrected_word_list_without_stopwords_lemmatized
corrected_word_list_without_stopwords_lemmatized = \
get_lemma(corrected_word_list_without_stopwords_stemmed)
corrected_word_list_without_stopwords_lemmatized[:20]
```

This code generates the following output:

```
['the',
 'reader',
 'cours',
 'basic',
 'knowledg',
 'python',
 'program',
 'languag',
 '.',
 'he/sh',
 'must',
 'knowledg',
 'data',
 'type',
 'python',
 '.',
 'He',
 'abl',
 'write',
 'function']
```

Figure 1.8: List of lemmatized words

In the preceding code, we looped through each of the words in the **corrected_word_list_without_stopwords** list and applied lemmatization to them. The preceding figure shows a list of the initial 20 lemmatized words.

13. To detect the sentence boundary in the given text corpus, use the **sent_tokenize()** method. Insert a new cell and add the following code to implement this:

```
print(sent_tokenize(corrected_sentence))
```

This code generates the following output:

```
[' The reader of this course should have a basic knowledge of the
Python programming language.', 'He/she must have knowledge of
data types in Python.', 'He should be able to write functions and
also have the ability to import and use libraries and packages in
Python.', 'familiarity with basic linguistics and probability is
assumed although not required to fully complete this course.']
```

> **NOTE**
>
> To access the source code for this specific section, please refer to https://packt.live/3gmyclC.
>
> You can also run this example online at https://packt.live/2D3h0ms.

CHAPTER 2: FEATURE EXTRACTION METHODS

ACTIVITY 2.01: EXTRACTING TOP KEYWORDS FROM THE NEWS ARTICLE

Solution

The following steps will help you complete this Activity:

1. Open a Jupyter Notebook.

2. Insert a new cell and add the following code to import the necessary libraries and download the data:

```
import operator

from nltk.tokenize import WhitespaceTokenizer
from nltk import download, stem

# The below statement will download the stop word list
# 'nltk_data/corpora/stopwords/' at home directory of your computer
download('stopwords')
from nltk.corpus import stopwords
```

The **download** statement will download the stop word list at **nltk_data/corpora/stopwords/** into your system's home directory.

3. Create the different types of methods to perform various NLP tasks:

`Activity 2.01.ipynb`

```
def load_file(file_path):
    news = ''.join\
            ([line for line in open(file_path,encoding='utf-8')])
    return news

"""
This method will take string as input and return the string
converted into lowercase
"""
def to_lower_case(text):
    return text.lower()

# This will take a text string as input and return the token.
wht = WhitespaceTokenizer()
def tokenize_text(text):
    return wht.tokenize(text=text)
```

The full code snippet can be found at https://packt.live/3hRl3kl

The **load_file()** function will take the file path as input and return the content of the file as a string. The **lower_case()** function will take a string as an argument and convert it into lowercase. Next, the **tokenize_text()** function will tokenize the string into its constituent tokens. The **get_stem()** method will perform stemming on the tokens, while **get_freq()** will calculate the frequency of the tokens. Finally, **get_top_n_words()** will return the n tokens with the highest frequency.

4. Load a text file into a string using the **load_file()** method:

```
path = "../data/news_article.txt"
news_article = load_file(path)
```

5. Convert the text into lowercase using the **to_lower_case()** method:

```
lower_case_news_art = to_lower_case(text=news_article)
```

6. Tokenize the text with the **tokenize_text()** method using the following line of code:

```
tokens = tokenize_text(lower_case_news_art)
```

7. Remove the stop words from the list; add the following code to do this:

```
removed_tokens = remove_stop_words(tokens)
```

8. Perform stemming on the words using the **get_stems()** method:

```
stems = get_stems(removed_tokens)
```

9. Now, calculate the frequency of stemmed tokens with the **get_freq()** method:

```
freq_dict = get_freq(stems)
```

10. To get the top six most frequently used words in the news article, use the following code:

```
top_keywords = get_top_n_words(freq_dict, 6)
top_keywords
```

The preceding line of code will generate the following output:

```
['law', 'justic', 'european', 'parti', 'took', 'poland'']
```

Thus, we have extracted the top six keywords from the news article, which can give us an idea of what the article is about. However, in this example, we have extracted only unigrams. For a more comprehensive output, bigrams and trigrams are often more useful. So, for even better results, you can perform the preceding activity on bigrams and trigrams.

> **NOTE**
>
> To access the source code for this specific section, please refer to https://packt.live/3hRl3kl.
>
> You can also run this example online at https://packt.live/2DnUHaU.

ACTIVITY 2.02: TEXT VISUALIZATION

Solution

1. Open a Jupyter Notebook. Insert a new cell and add the following code to import the necessary libraries:

```
from wordcloud import WordCloud, STOPWORDS
import matplotlib.pyplot as plt
%matplotlib inline
from nltk import word_tokenize
from nltk.stem import WordNetLemmatizer
import nltk
nltk.download('punkt')
from collections import Counter
import re
import matplotlib as mpl
mpl.rcParams['figure.dpi'] = 300
```

2. To fetch the dataset and read its content, add the following code:

```
text = open('../data//text_corpus.txt', 'r', \
            encoding='utf-8').read()
text[:1040]
```

The preceding code generates the following output:

```
'New Zealand players Jesse Tashkoff and Joey Field won hearts on Wednesday after they helped carry injured
West Indies batsman Kirk McKenzie off the field during their U-19 World Cup quarterfinal.\n\nMcKenzie was
suffering from cramps on his leg and was forced to retire hurt on 99, but courageously decided to return t
o face the final 14 deliveries of the innings after West Indies had lost their ninth wicket.\n\nAfter stru
ggling to hobble back to the middle, however, he was dismissed first ball -- still one short of his centur
y.\n\nCue the intervention of the sportsmanlike Kiwi duo, who carried the badly-cramped McKenzie off the f
ield while being applauded back to the pavilion by the fans at Benoni\'s Willowmoore Park in South Afric
a.\n\nThe gesture was applauded by the cricketing fraternity with India vice-captain Rohit Sharma also pra
ising the New Zealand players.\n\n"So good to see this #SpiritOfCricket at its best," Rohit tweeted along
with the video of the incident.\n\n@ImRo45\n So good to see this #SpiritOfCricket at its best. https:/'
```

Figure 2.31: Text corpus

3. The text in the fetched data is not clean. In order to clean it, we need to make use of various preprocessing steps, such as tokenization and lemmatization. Add the following code to implement this:

```
def lemmatize_and_clean(text):
    nltk.download('wordnet')
    lemmatizer = WordNetLemmatizer()
    cleaned_lemmatized_tokens = [lemmatizer.lemmatize\
                                 (word.lower()) \
                                 for word in word_tokenize\
                                 (re.sub(r'([^\s\w]|_)+', ' ', \
                                 text))]
    return cleaned_lemmatized_tokens
```

4. To check the set of unique words, along with their frequencies, as well as to find the 50 most frequently occurring words, add the following code:

```
Counter(lemmatize_and_clean(text)).most_common(50)
```

The preceding code generates the following output:

```
[('the', 31),
 ('to', 12),
 ('and', 8),
 ('of', 8),
 ('field', 7),
 ('on', 6),
 ('with', 5),
 ('new', 5),
 ('zealand', 5),
 ('in', 5),
 ('by', 4),
 ('wa', 4),
 ('s', 4),
 ('mckenzie', 4),
 ('clarke', 4),
 ('wicket', 4),
 ('at', 3),
 ('2', 3),
 ('joey', 3),
 ('spiritofcricket', 3),
 ('after', 3),
 ('unbeaten', 3),
 ('final', 3),
 ('indie', 3),
 ('west', 3),
 ('their', 3),
 ('it', 3),
 ('ball', 2),
 ('world', 2),
```

Figure 2.32: The 50 most frequent words

5. Once you get the set of unique words along with their frequencies, remove the stop words. Then, generate the word cloud for the top 50 most frequent words. Add the following code to implement this:

```
stopwords = set(STOPWORDS)
cleaned_text = ' '.join(lemmatize_and_clean(text))
wordcloud = WordCloud(width = 800, height = 800, \
                        background_color ='white', \
                        max_words=50, \
                        stopwords = stopwords, \
                        min_font_size = 10).generate(cleaned_text)
plt.imshow(wordcloud, interpolation='bilinear')
plt.axis("off")
plt.show()
```

The preceding code generates the following output:

Figure 2.33: Word cloud representation of the 50 most frequent words

As shown in the preceding image, words that occur more frequently, such as "unbeaten," "final," and "wicket," appear in larger sizes in the word cloud.

> **NOTE**
>
> To access the source code for this specific section, please refer to https://packt.live/30cDHxt.
>
> You can also run this example online at https://packt.live/33buXtj.

CHAPTER 3: DEVELOPING A TEXT CLASSIFIER

ACTIVITY 3.01: DEVELOPING END-TO-END TEXT CLASSIFIERS

Solution

The following steps will help you implement this activity:

1. Open a Jupyter Notebook.

2. Insert a new cell and add the following code to import the necessary packages:

```
import pandas as pd
import seaborn as sns
import matplotlib.pyplot as plt
%matplotlib inline
from nltk import word_tokenize
from nltk.corpus import stopwords
from nltk.stem import WordNetLemmatizer
from sklearn.feature_extraction.text import TfidfVectorizer
from sklearn.model_selection import train_test_split
import nltk
nltk.download('stopwords')
nltk.download('punkt')
nltk.download('wordnet')
import warnings
import string
import re
warnings.filterwarnings('ignore')
from sklearn.metrics import accuracy_score, roc_curve, \
classification_report, confusion_matrix, \
precision_recall_curve, auc
```

3. Read a data file. It has three columns: **is_political**, **headline**, and **short_description**. The **headline** column contains various news headlines, the **short_description** column contains an abstract of the article, and the **is_political** column indicates whether the article is about politics or not. Here, label 0 denotes that a headline is not political and label 1 denotes that the headline is political. Here, we will only use the **short_ description** column to train our model. Add the following code to do this:

```
data = pd.read_csv('data/news_political_dataset.csv')
data.sample(5)
```

The preceding code generates the following output:

	headline	short_description	is_political
59005	Aiguille Du Midi Offers Epic, Thrilling Views ...	Aiguille du Midi has long been a destination f...	0
29908	Madonna Speaks Out After Widely Panned Prince ...	Questlove also defended the singer's performance.	0
8228	The Poultry Lobby Wants Trump To Let Them Spee...	Labor and safety groups are urging the Agricul...	1
31786	See All The Winners From The MTV Movie Awards ...	Because the anticipation was killing you.	0
65216	10 Family Adventures In Latin America (PHOTOS)	Help herd sheep with the gauchos on the Patago...	0

Figure 3.67: Text data and labels stored as a DataFrame

4. Create a generic function for all the classifiers called **clf_model**. It takes four inputs: the type of model, the features of the training dataset, the labels of the training dataset, and the features of the validation dataset. It returns predicted labels, predicted probabilities, and the model it has been trained on. Add the following code to do this:

```
def clf_model(model_type, X_train, y_train, X_valid):
    model = model_type.fit(X_train,y_train)
    predicted_labels = model.predict(X_valid)
    predicted_probab = model.predict_proba(X_valid)[:,1]
    return [predicted_labels,predicted_probab, model]
```

5. Furthermore, another function is defined, called **model_evaluation**. It takes three inputs: actual values, predicted values, and predicted probabilities. It prints a confusion matrix, accuracy, f1-score, precision, recall scores, and the AUROC curve. It also plots the ROC curve:

```
def model_evaluation(actual_values, predicted_values, \
                     predicted_probabilities):
    cfn_mat = confusion_matrix(actual_values,predicted_values)
    print("confusion matrix: \n",cfn_mat)
    print("\naccuracy: ",accuracy_score\
                        (actual_values,predicted_values))
    print("\nclassification report: \n", \
          classification_report(actual_values,predicted_values))

    fpr,tpr,threshold=roc_curve(actual_values, \
                                predicted_probabilities)
    print('\nArea under ROC curve for validation set:', \
          auc(fpr,tpr))
    fig, ax = plt.subplots(figsize=(6,6))
```

```
    ax.plot(fpr,tpr,label='Validation set AUC')
    plt.xlabel('False Positive Rate')
    plt.ylabel('True Positive Rate')
    ax.legend(loc='best')
    plt.show()
```

6. Use a **lambda** function to extract tokens from each text in this **DataFrame** (called data), check whether any of these tokens are stop words, lemmatize them, and then concatenate them side by side. Use the **join** function to concatenate a list of words into a single sentence. After that, use the regular expression method (**re**) to replace anything other than letters, digits, and whitespaces with blank spaces. Add the following code to implement this:

```
lemmatizer = WordNetLemmatizer()
stop_words = stopwords.words('english')
stop_words = stop_words + list(string.printable)
data['cleaned_headline_text'] = data['short_description']\
                        .apply(lambda x : ' '.join\
                        ([lemmatizer.lemmatize\
                          (word.lower()) \
                        for word in word_tokenize\
                        (re.sub(r'([^\s\w]|_)+', ' ', \
                        str(x))) if word.lower() \
                        not in stop_words]))
```

7. Create a TFIDF matrix representation of these cleaned texts. Add the following code to do this:

```
MAX_FEATURES = 200
tfidf_model = TfidfVectorizer(max_features=MAX_FEATURES)
tfidf_df = pd.DataFrame(tfidf_model.fit_transform\
            (data['cleaned_headline_text']).todense())
tfidf_df.columns = sorted(tfidf_model.vocabulary_)
tfidf_df.head()
```

The preceding code generates the following output:

	10	administration	already	also	always	america	american	another	around	away	...	white	win	without	woman	wc
0	0.0	0.0	0.0	0.0	0.0	0.0	0.0	0.0	0.0	0.0	...	0.0	0.0	0.0	0.0	0.0
1	0.0	0.0	0.0	0.0	0.0	0.0	0.0	0.0	0.0	0.0	...	0.0	0.0	0.0	0.0	0.0
2	0.0	0.0	0.0	0.0	0.0	0.0	0.0	0.0	0.0	0.0	...	0.0	0.0	0.0	0.0	0.0
3	0.0	0.0	0.0	0.0	0.0	0.0	0.0	0.0	0.0	0.0	...	0.0	0.0	0.0	0.0	0.0
4	0.0	0.0	0.0	0.0	0.0	0.0	0.0	0.0	0.0	0.0	...	0.0	0.0	0.0	0.0	0.0

5 rows × 200 columns

Figure 3.68: TFIDF representation of the DataFrame

8. Use sklearn's **train_test_split** function to divide the dataset into training and validation sets. Add the following code to do this:

```
X_train, X_valid, y_train, y_valid = \
train_test_split(tfidf_df, data['is_political'], test_size=0.2, \
                 random_state=42,stratify = data['is_political'])
```

9. Train an XGBoost model using the **XGBClassifier()** function and evaluate it for the validation set. Add the following code to do this:

```
pip install xgboost
from xgboost import XGBClassifier
xgb_clf=XGBClassifier(n_estimators=10,learning_rate=0.05,\
                      max_depth=18,subsample=0.6,\
                      colsample_bytree= 0.6,\
                      reg_alpha= 10,seed=42)
results = clf_model(xgb_clf, X_train, y_train, X_valid)
model_evaluation(y_valid, results[0], results[1])
model_xgb = results[2]
```

The preceding code generates the following output:

```
confusion matrix:
 [[6862  491]
 [3498 3050]]

accuracy:  0.7130422271779009

classification report:
               precision    recall  f1-score   support

           0       0.66      0.93      0.77      7353
           1       0.86      0.47      0.60      6548

    accuracy                           0.71     13901
   macro avg       0.76      0.70      0.69     13901
weighted avg       0.76      0.71      0.69     13901


Area under ROC curve for validation set: 0.7499565605185604
```

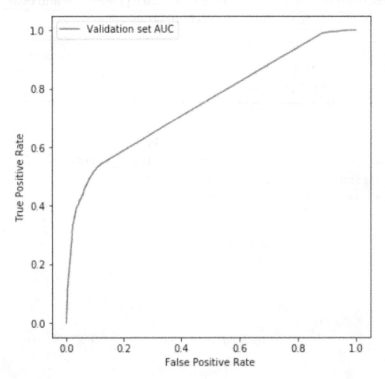

Figure 3.69: Performance of the XGBoost model

10. Extract the importance of features, that is, tokens or words that play a vital role in determining the type of content. Add the following code to do this:

```
word_importances = pd.DataFrame\
                    ({'word':X_train.columns,\
                      'importance':model_xgb.feature_importances_})
word_importances.sort_values('importance', \
                    ascending = False).head(4)
```

The preceding code generates the following output:

	word	importance
178	trump	0.192977
136	president	0.093427
144	republican	0.070339
25	clinton	0.043241

Figure 3.70: Words and their importance

NOTE

To access the source code for this specific section, please refer to https://packt.live/2ParIKD.

You can also run this example online at https://packt.live/33axiEK.

CHAPTER 4: COLLECTING TEXT DATA WITH WEB SCRAPING AND APIS

ACTIVITY 4.01: EXTRACTING INFORMATION FROM AN ONLINE HTML PAGE

Solution

Let's extract the data from an online source and analyze it. Follow these steps to implement this activity:

1. Open a Jupyter Notebook.

2. Import the **requests** and **BeautifulSoup** libraries. Pass the URL to **requests** with the following command. Convert the fetched content into HTML format using the **BeautifulSoup** HTML parser. Add the following code to do this:

```
import requests
from bs4 import BeautifulSoup
r = requests\
    .get('https://en.wikipedia.org/wiki/Rabindranath_Tagore')
r.status_code
soup = BeautifulSoup(r.text, 'html.parser')
```

3. To extract the list of headings, see which HTML elements belong to each bold headline in the *Works* section. You can see that they belong to the **h3** tag. We only need the first six headings here. Look for a **span** tag that has a **class** attribute with the following set of commands:

```
for element in soup.find_all('h3', limit=6):
    spans = element.find('span', attrs={'class':"mw-headline"})
    print(spans['id'])
```

The preceding code generates the following output:

```
Drama
Short_stories
Novels
Poetry
Songs_(Rabindra_Sangeet)
Art_works
```

4. To extract information regarding the original list of works written in Bengali by Tagore, look for the **table** tag. Traverse through the table and use **select** to pick table rows (**tr**) from following table data (**td**) associated with it. Add the following code to extract the text:

```
table = soup.find('table', attrs={'class':"wikitable"})
for row in table.select('tr td'):
    print(row.text)
```

The preceding code generates the following output:

* ভানুসিংহ ঠাকুরের পদাবলী
Bhānusiṃha Ṭhākurer Paḍāvalī
(Songs of Bhānusiṃha Ṭhākur)
1884

* মানসী
Manasi
(The Ideal One)
1890

* সোনার তরী
Sonar Tari
(The Golden Boat)
1894

* গীতাঞ্জলি
Gitanjali
(Song Offerings)
1910

Figure 4.16: List of Tagore's work

5. To extract the list of universities named after Tagore, look for the **ol** tag. Add the following code to do this:

```
[each.text.strip() for each in soup.find('ol') if each != '\n']
```

The preceding code generates the following output:

```
['Rabindra Bharati University, Kolkata, India.',
 'Rabindra University, Sahjadpur, Shirajganj, Bangladesh.[1]',
 'Rabindranath Tagore University, Hojai, Assam, India',
 'Rabindra Maitree University, Courtpara, Kustia,Bangladesh.[2]',
 'Bishwakabi Rabindranath Tagore Hall, Jahangirnagar University, Bangladesh',
 'Rabindra Nazrul Art Building, Arts Faculty, Islamic University, Bangladesh',
 'Rabindra Library (Central), Assam University, India',
 'Rabindra Srijonkala University, Keraniganj, Dhaka, Bangladesh']
```

Figure 4.17: List of universities named after Rabindranath Tagore

> **NOTE**
>
> To access the source code for this specific section, please refer to https://packt.live/315vOcd.
>
> You can also run this example online at https://packt.live/2D6qIV9.

ACTIVITY 4.02: EXTRACTING AND ANALYZING DATA USING REGULAR EXPRESSIONS

Solution

Follow these steps to complete this activity:

1. Collect the data using the **requests** package with the following code:

```
import requests
from bs4 import BeautifulSoup
r = requests.get('https://www.packtpub.com/support/faq')
r.status_code
```

The preceding code generates the following output:

```
200
```

2. Convert the fetched content into HTML format using **BeautifulSoup**'s HTML parser.

```
soup = BeautifulSoup(r.text, 'html.parser')
```

3. Inspect the HTML tag of the Packt website FAQs page. You can extract the question text by first searching for the **div** tag with the **"class":"tab"** attribute and inside that element, find the **label** tag to get the question text. Similarly, to get the answer text, find the **div** tag with **"class":"tab-content"**, as shown here:

```
qas = []
for each in soup.find_all('div', attrs={"class":"tab"}):
    question = each.find('label')
    answer = each.find('div', attrs={"class":"tab-content"})
    qas.append((question.text, answer.text))
print(qas[1])
```

The preceding code generates the following output:

```
('What format are Packt eBooks?', '\nPackt eBooks can be downloaded
as a PDF, EPUB or MOBI file. They can also be viewed online using your
subscription.\n')
```

4. Create a DataFrame consisting of these questions and answers:

```
import pandas as pd
pd.DataFrame(qas, columns=['Question', 'Answer']).head()
```

The preceding code generates the following output:

	Question	Answer
0	How can I download eBooks?	\nOnce you complete your eBook purchase, the d...
1	What format are Packt eBooks?	\nPackt eBooks can be downloaded as a PDF, EPU...
2	Can I send an eBook to my Kindle?	\nYes, if you follow the previous instructions...
3	How can I download code files for eBooks and V...	\nThere are a number of simple ways to access ...
4	How can I download Videos?	\nOnce you complete your Video purchase, the d...

Figure 4.18: DataFrame of the question and answers

5. To extract email addresses, make use of a regular expression. Insert a new cell and add the following code to implement this:

```
tc_page_r = requests\
            .get('https://www.packtpub.com/books/info/'\
            'packt/terms-and-conditions')
tc_page_r.status_code

soup2 = BeautifulSoup(tc_page_r.text, 'html.parser')

import re
set(re.findall\
    (r"[A-Za-z0-9._%+-]+@[A-Za-z0-9.-]+\.[A-Za-z]{2,4}",\
    soup2.text))
```

Here, the regular expression pattern will be looking for an alphanumeric blob, followed by the @ sign, followed by an alphanumeric blob. Next, it will look for a dot (.) followed by a 2 to 4-character suffix for domains (**com/in/org**). The preceding code generates the following output:

```
{'customercare@packt.com', 'subscription.support@packt.com'}
```

6. To extract phone numbers using a regular expression, insert a new cell and add the following code:

```
re.findall(r"\+\d{2}\s{1}\(0\)\s\d{3}\s\d{3}\s\d{3}", soup2.toxt)
```

The preceding code generates the following output:

```
['+44 (0) 121 265 648', '+44 (0) 121 212 141']
```

> **NOTE**
>
> To access the source code for this specific section, please refer to https://packt.live/2D4ijBK.
>
> You can also run this example online at https://packt.live/3jSXRVb.

ACTIVITY 4.03: EXTRACTING DATA FROM TWITTER

Solution

Let's extract tweets using the **tweepy** library. Follow these steps to implement this activity:

1. Log in to your Twitter account with your credentials.

2. Visit https://dev.twitter.com/apps/new, fill in the necessary details, and submit the form.

3. Once the form is submitted, go to the **Keys** and **tokens** tab; copy **consumer_key**, **consumer_secret**, **access_token**, and **access_token_secret** from there.

4. Open a Jupyter Notebook.

5. Import the relevant packages and follow the authentication steps by writing the following code:

```
consumer_key = 'your_consumer_key'
consumer_secret = 'your_consumer_secret'
access_token = 'your_access_token'
access_token_secret = 'your_access_token_secret'
import pandas as pd
import json
from pprint import pprint
import tweepy

auth = tweepy.OAuthHandler(consumer_key, consumer_secret)
auth.set_access_token(access_token, access_token_secret)

api = tweepy.API(auth)
```

6. Call the Twitter API with the **#climatechange** search query. Insert a new cell and add the following code to implement this:

```
tweet_list = []
for tweet in tweepy.Cursor(api.search, q='#climatechange', \
                           lang="en").items(100):
    tweet_list.append(tweet)
len(tweet_list)
tweet_list[0]
```

The preceding code generates an output that should look similar to the following screenshot. The content will vary since the tweets will be different according to when you are running the program:

```
Status(_api=<tweepy.api.API object at 0x11cb6f550>, _json={'created_at': 'Sun Jan 26 12:42:47 +0000 2020',
'id': 1221413171245801472, 'id_str': '1221413171245801472', 'text': 'The latest The Passivhaus Daily! http
s://t.co/sPqQhgSRdo Thanks to @TheMarkofPolo @PeterGleick @boris_kapkov #passivehouse #climatechange', 'tr
uncated': False, 'entities': {'hashtags': [{'text': 'passivehouse', 'indices': [109, 122]}, {'text': 'clim
atechange', 'indices': [123, 137]}], 'symbols': [], 'user_mentions': [{'screen_name': 'TheMarkofPolo', 'na
me': 'Mark Ostendorf', 'id': 37373892, 'id_str': '37373892', 'indices': [67, 81]}, {'screen_name': 'PeterG
leick', 'name': 'Peter Gleick', 'id': 146123790, 'id_str': '146123790', 'indices': [82, 94]}, {'screen_nam
e': 'boris_kapkov', 'name': 'Boris Kapkov', 'id': 487757126, 'id_str': '487757126', 'indices': [95, 10
8]}], 'urls': [{'url': 'https://t.co/sPqQhgSRdo', 'expanded_url': 'https://paper.li/greenhhome/green-passi
vehouse-design?edition_id=59bfd880-4039-11ea-ae35-0cc47a0d15fd', 'display_url': 'paper.li/greenhhome/gr
e…', 'indices': [33, 56]}]}, 'metadata': {'iso_language_code': 'en', 'result_type': 'recent'}, 'source':
```

Figure 4.19: The Twitter API called with the #climatechange search query

7. Each **tweepy Status** object will have a **json** object associated with it, which will have tweet content and meta info. Let's see what information is present:

```
status = tweet_list[0]
status_json = status._json
pprint(status_json)
```

The preceding code generates the following output with different tweet content fetched at the time of running the program:

```
{'contributors': None,
 'coordinates': None,
 'created_at': 'Sun Jan 26 12:42:47 +0000 2020',
 'entities': {'hashtags': [{'indices': [109, 122], 'text': 'passivehouse'},
                           {'indices': [123, 137], 'text': 'climatechange'}],
              'symbols': [],
              'urls': [{'display_url': 'paper.li/greenhhome/gre…',
                        'expanded_url': 'https://paper.li/greenhhome/green-passivehouse-design?edition_id=
59bfd880-4039-11ea-ae35-0cc47a0d15fd',
                        'indices': [33, 56],
                        'url': 'https://t.co/sPqQhgSRdo'}],
              'user_mentions': [{'id': 37373892,
                                 'id_str': '37373892',
                                 'indices': [67, 81],
                                 'name': 'Mark Ostendorf',
                                 'screen_name': 'TheMarkofPolo'},
                                {'id': 146123790,
                                 'id_str': '146123790',
                                 'indices': [82, 94],
                                 'name': 'Peter Gleick',
                                 'screen_name': 'PeterGleick'},
                                {'id': 487757126,
                                 'id_str': '487757126',
                                 'indices': [95, 108],
```

Figure 4.20: Twitter status objects converted to JSON objects

8. To check the tweet text, use the following code:

```
status_json['text']
```

Again, though the content may vary, the preceding code generates output similar to the following:

```
'The latest The Passivhaus Daily! https://t.co/sPqQhgSRdo Thanks
to @TheMarkofPolo @PeterGleick @boris_kapkov #passivehouse
#climatechange'
```

9. To create a DataFrame consisting of the text of tweets, add a new cell and write the following code:

```
tweets = []
for twt in tweet_list:
    tweets.append(twt._json['text'])


tweet_text_df = pd.DataFrame({'tweet_text' : tweets})
tweet_text_df.head()
```

The preceding code generates the following output. Again, the content may vary depending on the current tweets:

	tweet_text
0	The latest The Passivhaus Daily! https://t.co/...
1	RT @GabeFilippelli: Disgusting abuse of power,...
2	RT @Nornenland: #Australia dust storms are #so...

Figure 4.21: DataFrame with the text of tweets

> **NOTE**
>
> To access the source code for this specific section, please refer to https://packt.live/3jXyx03.
>
> This section does not currently have an online interactive example and will need to be run locally.

CHAPTER 5: TOPIC MODELING

ACTIVITY 5.01: TOPIC-MODELING JEOPARDY QUESTIONS

Solution

Let's perform topic modeling on the dataset of Jeopardy questions:

1. Open a Jupyter Notebook.

2. Insert a new cell and add the following code to import pandas and other libraries:

```
import numpy as np
import spacy
nlp = spacy.load('en_core_web_sm')
import pandas as pd
pd.set_option('display.max_colwidth', 800)
```

3. After downloading the data, you can extract it and place at the location below. Then load the Jeopardy CSV file into a pandas DataFrame. Insert a new cell and add the following code:

```
JEOPARDY_CSV =  '../data/jeopardy/Jeopardy.csv'
questions = pd.read_csv(JEOPARDY_CSV)
questions.columns = [x.strip() for x in questions.columns]
```

4. The data in the DataFrame is not clean. In order to clean it, remove records that have missing values in the **Question** column. Add the following code to do this:

```
questions = questions.dropna(subset=['Question'])
```

5. Find the number of unique categories. Add the following code to do this:

```
questions['Category'].nunique()
```

The code generates the following output:

```
27995
```

6. Sample 4% of the questions and tokenize the corpus where the tokens are classified as **NOUN** by spaCy:

```
file='../data/JQuestions.txt'
questions['Question'].sample(frac=0.04,replace=False,\
                            random_state=0).to_csv(file)


f=open(file,'r',encoding='utf-8')
text=f.read()
f.close()

doc=nlp(text)
pos_list=['NOUN']
preproc_text=[]
preproc_sent=[]

for token in doc:
    if token.text!='\n':
        if not(token.is_stop) and not(token.is_punct) \
        and token.pos_ in pos_list:
            preproc_sent.append(token.lemma_)
    else:
        preproc_text.append(preproc_sent)
        preproc_sent=[]

preproc_text.append(preproc_sent) #last sentence

print(preproc_text)
```

The code generates output like the following:

```
[['question'], ['king', 'lot'], ['grad', 'country'], ['leader'], ['woman', 'work'], ['senator'], ['hit',
'1'], ['slogan', 'estate', 'company', 'home', 'minute'], ['magazine', 'publisher'], ['mystery', 'work'],
['href=""http://www.j', 'media/2008', 'actress'], ['company', 'jean', 'mall'], ['array', 'row', 'column'],
['blood', 'toil', 'tear', 'horse', 'race', 'track'], ['href=""http://www.j', 'man', 'body', 'water'], ['ad
ult', 'human', 'capacity', 'liter'], ['href=""http://www.j', 'media/1990', '1st', 'year', 'girl', 'boy',
'wagon', '""</i'], [], [], ['a.m.', 'p.m.', 'capital'], ['stage', 'manager', 'presentation', 'drama'], ['p
icture', 'wall', 'sensibility'], ['signing', 'place', 'city'], ['composer', 'music', 'instrument'], ['roc
k', 'singer', 'box'], ['creature'], ['navy', 'man', 'island', 'sound'], ['day', 'subtitle', 'movie', 'sequ
```

Figure 5.22: Tokenized corpus after selecting 4% of the sample

7. Train a tomotopy LDA model with 1,000 topics. Print a few topics. Add the following code to do this:

```
import tomotopy as tp
NUM_TOPICS=1000


mdl = tp.LDAModel(k=NUM_TOPICS,seed=1234)


for line in preproc_text:
    mdl.add_doc(line)


mdl.train(10)


for k in range(mdl.k):
    print('Top 7 words of topic #{}'.format(k))
    print(mdl.get_topic_words(k, top_n=7))
```

The code generates the following output:

```
Top 7 words of topic #0
[('edition', 0.046412043273448944), ('type', 0.03483796492218971), ('city', 0.011689815670251846), ('web',
0.011689815670251846), ('country', 0.011689815670251846), ('creature', 0.011689815670251846), ('age', 0.01
1689815670251846)]
Top 7 words of topic #1
[('state', 0.08959732204675674), ('square', 0.0224832221865654), ('wood', 0.0224832221865654), ('plate',
0.0224832221865654), ('moon', 0.011297539807856083), ('o', 0.011297539807856083), ('hotel', 0.011297539807
856083)]
```

Figure 5.23: Topics inferred after training the LDA model

8. Now print the log perplexity. Add the following code to do this:

```
print('Log perplexity=',mdl.ll_per_word)
```

The code generates output like so:

```
Log perplexity= -14.396450040387437
```

9. Insert a new cell and add the following code to see the probability distribution of topics if we consider the entire dataset as a single document:

```
bag_of_words=[word for sent in preproc_text for word in sent]
doc_inst = mdl.make_doc(bag_of_words)
np.argsort(np.array(mdl.infer(doc_inst)[0]))[::-1]
```

The code generates output like so:

```
array([461, 234, 640, 679, 852, 978, 686, 862, 844, 748, 450, 504, 377,
       404, 992, 310, 840, 128, 565, 827, 899, 109, 751, 367, 782, 559,
       428, 996,  93, 684, 163, 121, 731, 519, 922, 639, 726, 645,  91,
       320, 405, 270, 815, 786, 126, 500, 512, 944, 113, 929, 412, 765,
       954, 553, 397, 332, 671, 498, 875, 792,  68, 296, 974, 592, 236,
        50, 798, 237, 140, 704, 646, 133, 456, 744, 204, 244, 691, 550,
       773, 607, 242, 386, 695, 104, 957,  47, 543, 895,  81, 724, 960,
```

Figure 5.24: Probability distribution of topics if the entire dataset is considered

10. Insert a new cell and add the following code to see the probability distribution of topic 461:

```
print(mdl.get_topic_words(461, top_n=7))
```

The code generates output like so:

```
[('city', 0.15946216881275177), ('device', 0.02001992054283619),
 ('force', 0.02001992054283619), ('character', 0.2001992054283619),
 ('death', 0.010059761814773083), ('person', 0.010059761814773083),
 ('language', 0.010059761814773083)]
```

11. Insert a new cell and add the following code to see the probability distribution of topic 234:

```
print(mdl.get_topic_words(234, top_n=7))
```

The code generates output like so:

```
[('year', 0.09871795773506165), ('group', 0.02968442067503929),
 ('child', 0.019822485744953156), ('murder', 0.019822485744953156),
 ('field', 0.019822485744953156), ('writing', 0.009960552677512169),
 ('memorial', 0.009960552677512169)]
```

12. Insert a new cell and add the following code to see the probability distribution of topic 186:

```
print(mdl.get_topic_words(186, top_n=7))
```

The code generates output like so:

```
[('dragon', 0.027016131207346916), ('power', 0.01357526984065711),
 ('flying', 0.013575269840657711), ('line', 0.013575269840657711),
 ('process', 0.013575269840657711),
 ('crystal', 0.013575269840657711),
 ('freestyle', 0.013575269840657711)]
```

We find that the log perplexity is around -14, but the topics are not interpretable and the number of categories is an order of magnitude greater than the number of topics. The topic model could still be used for dimensionality reduction.

> **NOTE**
>
> In general, the topics found are extremely sensitive to randomization in both gensim and tomotopy. While setting a **random_state** in gensim could help with reproducibility, in general, the topics found using tomotopy are superior from the perspective of interpretability. Generally, your output is expected to be different. In order to have exactly the same topic model, we can save and load topic models, and we do this in *Exercise 5.04, Topics in The Life and Adventures of Robinson Crusoe by Daniel Defoe*.
>
> To access the source code for this specific section, please refer to https://packt.live/33c2O5p.
>
> This section does not currently have an online interactive example, and will need to be run locally.

ACTIVITY 5.02: COMPARING DIFFERENT TOPIC MODELS

Solution

Let's perform topic modeling on the CFPB dataset. Follow these steps to complete this activity:

1. Open a Jupyter Notebook.

2. Insert a new cell and add the following code to import the pandas library:

```
import numpy as np
import spacy
nlp = spacy.load('en_core_web_sm')
file_student='../data/consumercomplaints/'\
                'student_comp_narrative.txt'
f=open(file_student,'r',encoding='utf-8')
student_text=f.read()
f.close()
```

3. Tokenize and include only nouns:

```
doc_student=nlp(student_text)
student_pos_list=['NOUN']
student_preproc_text=[]
student_preproc_sent=[]

for token in doc_student:
    if token.text!='\n':
        if not(token.is_stop) and not(token.is_punct) \
        and token.pos_ in student_pos_list:
            student_preproc_sent.append(token.lemma_)
    else:
        student_preproc_text.append(student_preproc_sent)
        student_preproc_sent=[]

student_preproc_text.append(student_preproc_sent) #last sentence

print(student_preproc_text)
```

The code generates the following output:

```
[['xx', 'payment', 'credit', 'union', 'paper', 'check', 'payment', 'copy', 'statement', 'payment', 'paymen
t', 'payment', 'payment', 'penalty', 'fee', 'foresearance', 'tooth', 'word'], ['student', 'loan', 'compan
y', 'enrollment', 'school', 'program', 'month', 'email', 'school', 'enrollment', 'information', 'enrollmen
t', 'loan', 'school', 'fee', 'hardship'], ['9259,i', 'call', 'differ', 't', 'amount'], ['deferment', 'requ
est', 'credit', 'bureau', 'company', 'report', 'impression', 'payment', 'report', 'day', 'total', 'accoun
t', 'standard', 'servicing', 'admission', 'company', '3year'], ['contract', 'loan', 'deferment', 'year',
```

Figure 5.25: Tokenized corpus containing only nouns

4. Train an HDP model and print the log perplexity and topics:

```
import tomotopy as tp
mdl = tp.HDPModel(alpha=0.1,seed=0)

for line in student_preproc_text:
    mdl.add_doc(line)

mdl.train(50)
print('Log Perplexity=', mdl.ll_per_word)

for k in range(mdl.k):
    print('Top 10 words of topic #{}'.format(k))
    print(mdl.get_topic_words(k, top_n=10))
```

The code generates the following output:

```
Log Perplexity= -5.74896161113102
Top 10 words of topic #0
[('payment', 0.07293784618377686), ('loan', 0.07257163524627686), ('account', 0.02941964566707611), ('tim
e', 0.024841995909810066), ('month', 0.02063055709004402), ('credit', 0.017945000901818275), ('informatio
n', 0.017883965745568275), ('interest', 0.015991870313882828), ('student', 0.015259445644915104), ('day',
0.014099773950874805)]
```

Figure 5.26: Log perplexity and the topics inferred from the HDP model

5. Insert a new cell and add the following code to save the topic model:

```
mdl.save('../data/consumercomplaints/hdp_model.bin')
```

6. Insert a new cell and add the following code to load the topic model:

```
mdl = tp.HDPModel.load('../data/consumercomplaints/hdp_model.bin')
```

7. Insert a new cell and add the following code to see the probability distribution of topics if we consider the entire dataset as a single document:

```
bag_of_words=[word for sent in student_preproc_text \
                for word in sent]
doc_inst = mdl.make_doc(bag_of_words)
np.argsort(np.array(mdl.infer(doc_inst)[0]))[::-1]
```

The code generates the following output:

```
array([5, 7, 4, 6, 0, 1, 11, 9, 14, 2, 18, 17, 10, 8, 12,
       13, 15, 3, 16], dtype=int64)
```

8. Insert a new cell and add the following code to see the probability distribution of topic 5:

```
print(mdl.get_topic_words(5, top_n=7))
```

The code generates the following output:

```
[('school', 0.05379803851246834), ('aid', 0.05379803851246834),
 ('password', 0.003592493385076523),
 ('username', 0.03592493385076523),
 ('information', 0.03592493385076523),
 ('direction', 0.03592493385076523), ('bus', 0.03592493385076523)]
```

9. Insert a new cell and add the following code to see the probability distribution of topic 7:

```
print(mdl.get_topic_words(7, top_n=7))
```

The code generates the following output:

```
[('graduate', 0.061739806085824966),
 ('program', 0.061739806085824966),
 ('assistance', 0.04634334146976471),
 ('loan', 0.03094688430428505), ('school', 0.03094688430428505),
 ('world', 0.03094688430428505)]
```

10. Insert a new cell and add the following code to see the probability distribution of topic 4:

```
print(mdl.get_topic_words(4, top_n=7))
```

The code generates the following output:

```
[('employer', 0.03343059867620468),
 ('graduation', 0.03343059867620468),
 ('book', 0.03343059867620468),
 ('diploma', 0.025093790143728256),
 ('debt', 0.025093790143728256),
 ('education', 0.025093790143728256),
 ('college', 0.025093790143728256)]
```

11. Now, train the LDA model. Add the following code for this:

```
NUM_TOPICS=20

mdl = tp.LDAModel(k=NUM_TOPICS,alpha=0.1,seed=0)

for line in student_preproc_text:
    mdl.add_doc(line)

mdl.train(50)
print('Log Perplexity=', mdl.ll_per_word)

for k in range(mdl.k):
    print('Top 10 words of topic #{}'.format(k))
    print(mdl.get_topic_words(k, top_n=10))
```

The code generates the following output:

```
Log Perplexity= -6.423249988462239
Top 10 words of topic #0
[('forbearance', 0.1309245228767395), ('time', 0.08555688709020615), ('month', 0.07453903555870056), ('pay
ment', 0.03630059212446213), ('situation', 0.033708155155181885), ('income', 0.03111572004854679), ('yea
r', 0.024634627625346184), ('option', 0.022690299898386), ('paperwork', 0.018801646307110786), ('end', 0.0
16857318580150604)]
Top 10 words of topic #1
[('credit', 0.14882682263851166), ('loan', 0.0883079543709755), ('score', 0.059536680579185486), ('accoun
t', 0.04961555451154709), ('report', 0.03671808913350105), ('letter', 0.033741749823093414), ('status', 0.
029773302376270294), ('customer', 0.021836400032043457), ('day', 0.018860062584280968), ('service', 0.0168
7583699822426)]
```

Figure 5.27: Log perplexity and topics inferred from the LDA model

12. Insert a new cell and add the following code to save the topic model:

```
mdl.save('../data/consumercomplaints/lda_model.bin')
```

13. Insert a new cell and add the following code to load the topic model:

```
mdl = tp.LDAModel.load('../data/consumercomplaints/lda_model.bin')
```

14. Insert a new cell and add the following code to see the probability distribution of topics if we consider the entire dataset as a single document:

```
bag_of_words=[word for sent in preproc_text for word in sent]
doc_inst = mdl.make_doc(bag_of_words)
np.argsort(np.array(mdl.infer(doc_inst)[0]))[::-1]
```

The code generates the following output:

```
array([17,  7,  6,  8, 12,  0,  2,  4, 10,  5, 18, 14, 13, 11,
       16, 15,  9,  3,  1, 19], dtype=int64)
```

15. Insert a new cell and add the following code to see the probability distribution of topic 17:

```
print(mdl.get_topic_words(17, top_n=7))
```

The code generates the following output:

```
[('interest', 0.20065094530582428), ('loan', 0.16345429420471191),
 ('payment', 0.152724489569664), ('rate', 0.07046262919902802),
 ('balance', 0.04184982180595398), ('year', 0.0314776748418808),
 ('principal', 0.25755111128091812)]
```

16. Insert a new cell and add the following code to see the probability distribution of topic 7:

```
print(mdl.get_topic_words(7, top_n=7))
```

The code generates the following output:

```
[('loan', 0.14698922634124756), ('year', 0.09230735898017883),
 ('repayment', 0.08487062156200409),
 ('payment', 0.08312080055475235), ('plan', 0.07349679619073868),
 ('income', 0.05074914172291756), ('month', 0.03981276974081993)]
```

17. Insert a new cell and add the following code to see the probability distribution of topic 6:

```
print(mdl.get_topic_words(6, top_n=7))
```

The code generates the following output:

```
[('loan', 0.24387744069099426), ('time', 0.06379450112581253),
 ('student', 0.051527272909879684), ('m', 0.05103658139705659),
 ('money', 0.04514831304550171), ('payment', 0.03239039331674576),
 ('collection', 0.02794190190434456)]
```

For our dataset and with the experimentation undertaken, the LDA topics were much more interpretable than the HDP topics. As seen from the preceding outputs, the log perplexity of the LDA model is also better than the log perplexity of the HDP model. We did, of course, benefit from using the number of topics found by the HDP model when training the LDP model, and so this is not an entirely fair comparison. Rather, this illustrates that there could be benefits to using an HDP model first even if we later select the LDA model for better interpretability or better log perplexity.

> **NOTE**
>
> In general, the topics found are extremely sensitive to randomization in both gensim and tomotopy. While setting a **random_state** in gensim could help reproducibility, in general, the topics found using tomotopy are superior from the perspective of interpretability. Generally, your output is expected to be different. In order to have exactly the same topic model, we can save and load topic models, and this was used in *Exercise 5.04*, *Topics in The Life and Adventures of Robinson Crusoe by Daniel Defoe*.
>
> To access the source code for this specific section, please refer to https://packt.live/312B9Bf..

CHAPTER 6: VECTOR REPRESENTATION

ACTIVITY 6.01: FINDING SIMILAR NEWS ARTICLE USING DOCUMENT VECTORS

Solution

Follow these steps to complete this activity:

1. Open a Jupyter Notebook. Insert a new cell and add the following code to import all necessary libraries:

```
import warnings
warnings.filterwarnings("ignore")


from gensim.models import Doc2Vec
import pandas as pd
from gensim.parsing.preprocessing import preprocess_string, \
remove_stopwords
```

2. Now load the **news_lines** file.

```
news_file = '../data/sample_news_data.txt'
```

3. After that, you need to iterate over each headline in the file and split the columns, then create a DataFrame containing the headlines. Insert a new cell and add the following code to implement this:

```
with open(news_file, encoding="utf8", errors='ignore') as f:
    news_lines = [line for line in f.readlines()]

lines_df = pd.DataFrame()
indices  = list(range(len(news_lines)))
lines_df['news'] = news_lines
lines_df['index'] = indices

lines_df.head()
```

The code produces the following output:

	news	index
0	Top of the Pops leaves BBC One The BBC flagshi...	0
1	Oscars race enters final furlong The race for ...	1
2	US TV special for tsunami relief A US televisi...	2
3	Williamson lauds bowlers for adapting to atypi...	3
4	Housewives lift Channel ratings The debut of U...	4

Figure 6.27: Head of the DataFrame

4. You already have a trained document model named **docVecModel.d2v** in the previous exercise. Now you can simply load and use it. Insert a new cell and add the following code to implement this:

```
docVecModel = Doc2Vec.load('../data/docVecModel.d2v')
```

5. Now, since you have loaded the document model, create two functions, namely, **to_vector()** and **similar_news_articles()**. The **to_vector()** function converts the sentences into vectors. The second function, **similar_news_articles()**, implements the similarity check. It uses the **docVecModel.docvecs.most_similar()** function, which compares the vector against all the other lines it was built with. To implement this, insert a new cell and add the following code:

```
from gensim.parsing.preprocessing import preprocess_string, \
remove_stopwords

def to_vector(sentence):
    cleaned = preprocess_string(sentence)
    docVector = docVecModel.infer_vector(cleaned)
    return docVector
```

```
def similar_news_articles(sentence):
    vector = to_vector(sentence)
    similar_vectors = docVecModel.docvecs.most_similar\
                      (positive=[vector])
    print(similar_vectors)
    similar_lines = lines_df\
                    [lines_df.index==similar_vectors[0][0]].news
    return similar_lines
```

6. Now that you have created the functions, it is time to test them. Insert a new cell and add the following code to implement this:

```
similar_news_articles("US raise TV indecency US politicians "\
                      "are proposing a tough new law aimed at "\
                      "cracking down on indecency")
```

The code will generate the following output:

```
1958          Clarke to unveil immigration plan New controls
Name: news, dtype: object
```

> **NOTE**
>
> To access the source code for this specific section, please refer to https://packt.live/3hMOcgO.
>
> You can also run this example online at https://packt.live/3gbDFvg.

CHAPTER 7: TEXT GENERATION AND SUMMARIZATION

ACTIVITY 7.01: SUMMARIZING COMPLAINTS IN THE CONSUMER FINANCIAL PROTECTION BUREAU DATASET

Solution

Follow these steps to complete this activity:

1. Open a Jupyter Notebook and insert a new cell. Add the following code to import the required libraries:

```
import warnings
warnings.filterwarnings('ignore')
import os
import csv
import pandas as pd
from gensim.summarization import summarize
```

2. Insert a new cell and add the following code to fetch the Consumer Complaints dataset and consider the rows that have a complaint narrative. Drop all the columns other than **Product**, **Sub-product**, **Issue**, **Sub-issue**, and **Consumer complaint narrative**:

```
complaints_pathname = '../data/consumercomplaints/'\
                      'Consumer_Complaints.csv'
df_all_complaints = pd.read_csv(complaints_pathname)
df_all_narr = df_all_complaints.dropna\
            (subset=['Consumer complaint narrative'])
df_all_narr = df_all_narr[['Product','Sub-product','Issue',\
                          'Sub-issue',\
                          'Consumer complaint narrative']]
```

3. Insert a new cell and add the following code to select 12 complaints:

```
df_part_narr = df_all_narr[df_all_narr.index.isin\
            ([242830,1086741,536367,957355,975181,483530,\
            950006,865088,681842,536367,132345,285894])]
df_part_narr
```

The preceding code generates the following output:

	Product	Sub-product	Issue	Sub-issue	Consumer complaint narrative
132345	Mortgage	FHA mortgage	Loan servicing, payments, escrow account	NaN	I have a trial period of loan modification HAR...
242830	Debt collection	Medical	Cont'd attempts collect debt not owed	Debt is not mine	XXXX XX/XX/XXXX XXXX XXXX, XXXX As a 100 % rat...
285894	Prepaid card	General purpose card	Managing, opening, or closing account	NaN	I purchased a prepaid credit card from XXXX wh...
483530	Debt collection	I do not know	Cont'd attempts collect debt not owed	Debt is not mine	This company has called me XXXX times about a ...

Figure 7.19: DataFrame showing the 12 selected complaints

4. Insert a new cell and add the following code to add a new column, named **TextRank Summary**, that includes a TextRank summary for each of the 12 complaints:

```
def try_summarize(x,ratio):
    try:
        return(summarize(x,ratio=ratio))
    except:
        return('')
df_part_narr['TextRank Summary']=df_part_narr\
                        ['Consumer complaint narrative']\
                        .apply(lambda x: try_summarize\
                        (x,ratio=0.20))
```

5. Insert a new cell and add the following code to show the DataFrame:

```
df_part_narr
```

The preceding code generates the following output:

	Product	Sub-product	Issue	Sub-issue	Consumer complaint narrative	TextRank Summary
132345	Mortgage	FHA mortgage	Loan servicing, payments, escrow account	NaN	I have a trial period of loan modification HAR...	They will not acknowledge my modification, and...
242830	Debt collection	Medical	Cont'd attempts collect debt not owed	Debt is not mine	XXXX XX/XX/XXXX XXXX XXXX, XXXX As a 100 % rat...	XXXX Care coordinator XXXX XXXX contacted the ...
285894	Prepaid card	General purpose card	Managing, opening, or closing account	NaN	I purchased a prepaid credit card from XXXX wh...	Was refused and told I have to use the card.
483530	Debt collection	I do not know	Cont'd attempts collect debt not owed	Debt is not mine	This company has called me XXXX times about a ...	

Figure 7.20: DataFrame showing the summarized complaints

> **NOTE**
>
> To access the source code for this specific section, please refer to https://packt.live/313r5YP.
>
> This section does not currently have an online interactive example, and will need to be run locally.

CHAPTER 8: SENTIMENT ANALYSIS

ACTIVITY 8.01: TWEET SENTIMENT ANALYSIS USING THE TEXTBLOB LIBRARY

Solution

To perform sentiment analysis on the given set of tweets related to airlines, follow these steps:

1. Open a Jupyter Notebook.

2. Insert a new cell and add the following code to import the necessary libraries:

```
import pandas as pd
from textblob import TextBlob
import re
```

3. Since we are displaying the text in the notebook, we want to increase the display width for our DataFrame. Insert a new cell and add the following code to implement this:

```
pd.set_option('display.max_colwidth', 240)
```

4. Now, load the **airline-tweets.csv** dataset. We will read this CSV file using pandas' **read_csv()** function. Insert a new cell and add the following code to implement this:

```
tweets = pd.read_csv('data/airline-tweets.csv')
```

5. Insert a new cell and add the following code to view the first 10 records of the DataFrame:

```
tweets.head()
```

The code generates the following output:

	tweet
0	@FlyHighAirways What @dhepburn said.
1	@FlyHighAirways plus you've added commercials to the experience... tacky.
2	@FlyHighAirways I didn't today... Must mean I need to take another trip!
3	@FlyHighAirways it's really aggressive to blast obnoxious "entertainment" in your guests' faces & they have little recourse
4	@FlyHighAirways and it's a really big bad thing about it

Figure 8.8: The first few tweets

6. If we look at the preceding figure, we can see that the tweets contain Twitter handles, which start with the @ symbol. It might be useful to extract those handles. The **string** column included in the DataFrame has an **extract()** function, which uses a regex to get parts of a string. Insert a new cell and add the following code to implement this:

```
tweets['At'] = tweets['tweet'].str.extract(r'^(@\S+)')
```

This code declares a new column called **At** and sets the value to what the **extract** function returns. The **extract** function uses a regex, **^(@\S+)**, to return strings that start with @. To view the initial 10 records of the **tweets** DataFrame, we insert a new cell and write the following code:

```
tweets.head(10)
```

The output should look something like this (only top four tweets are shown here):

	tweet	At
0	@FlyHighAirways What @dhepburn said.	@FlyHighAirways
1	@FlyHighAirways plus you've added commercials to the experience... tacky.	@FlyHighAirways
2	@FlyHighAirways I didn't today... Must mean I need to take another trip!	@FlyHighAirways
3	@FlyHighAirways it's really aggressive to blast obnoxious "entertainment" in your guests' faces & they have little recourse	@FlyHighAirways

Figure 8.9: The first 10 tweets along with the Twitter handles

7. Now, we want to remove the Twitter handles inside the tweets since they are irrelevant for sentiment analysis. First, create a function named **remove_ handles()**, which accepts a DataFrame as a parameter. After passing the DataFrame, the **re.sub()** function will remove the handles in the DataFrame. Insert a new cell and add the following code to implement this:

```
def remove_handles(tweet):
    return re.sub(r'@\S+', '', tweet)
```

8. To remove the handles, insert a column in the DataFrame called **tweets_ preprocessed** and add the following code:

```
tweets['tweet_preprocessed'] = tweets['tweet']\
                               .apply(remove_handles)
tweets.head(10)
```

The expected output for the tweets after removing the Twitter handles should look like this (only the top four are shown in this figure):

	tweet	At	tweet_preprocessed
0	@FlyHighAirways What @dhepburn said.	@FlyHighAirways	What said.
1	@FlyHighAirways plus you've added commercials to the experience... tacky.	@FlyHighAirways	plus you've added commercials to the experience... tacky.
2	@FlyHighAirways I didn't today... Must mean I need to take another trip!	@FlyHighAirways	I didn't today... Must mean I need to take another trip!
3	@FlyHighAirways it's really aggressive to blast obnoxious "entertainment" in your guests' faces & they have little recourse	@FlyHighAirways	it's really aggressive to blast obnoxious "entertainment" in your guests' faces & they have little recourse

Figure 8.10: The first 10 tweets after removing the Twitter handles

From the preceding figure, we can see that the Twitter handles have been separated from the tweets.

9. Now we can apply sentiment analysis on the tweets. First, create a **get_sentiment()** function, which accepts a DataFrame and a column as parameters. Using this function, we create two new columns, **Polarity** and **Subjectivity**, which will show the sentiment scores of each tweet. Insert a new cell and add the following code to implement this:

```
def get_sentiment(dataframe, column):
    text_column = dataframe[column]
    textblob_sentiment = text_column.apply(TextBlob)
    sentiment_values = [{'Polarity': v.sentiment.polarity, \
                         'Subjectivity': v.sentiment.subjectivity}
                for v in textblob_sentiment.values]
    return pd.DataFrame(sentiment_values)
```

This function takes a DataFrame and applies the **TextBlob** constructor to each value of **text_column**. Then it extracts and creates a new DataFrame with the **Polarity** and **Objectivity** columns.

10. Since the function has been created, we test it and pass the necessary parameters. The result of this will be stored in a new DataFrame, **sentiment_frame**. Insert a new cell and add the following code to implement this:

```
sentiment_frame = get_sentiment(tweets, 'tweet_preprocessed')
```

11. To view the initial four values of the new DataFrame, type the following code:

```
sentiment_frame.head(4)
```

The code generates the following output:

	Polarity	Subjectivity
0	0.000000	0.0000
1	0.000000	0.0000
2	-0.390625	0.6875
3	0.006250	0.3500

Figure 8.11: Polarity and subjectivity scores

12. To join the original **tweet** DataFrame to the **sentiment_frame** DataFrame, use the **concat()** function. Insert a new cell and add the following code to implement this:

```
tweets = pd.concat([tweets, sentiment_frame], axis=1)
```

13. To view the initial 10 rows of the new DataFrame, we add the following code:

```
tweets.head(10)
```

The expected output with sentiment scores added should be as follows:

	tweet	At	tweet_preprocessed	Polarity	Subjectivity
0	@FlyHighAirways What @dhepburn said.	@FlyHighAirways	What said.	0.000000	0.000000
1	@FlyHighAirways plus you've added commercials to the experience... tacky.	@FlyHighAirways	plus you've added commercials to the experience... tacky.	0.000000	0.000000
2	@FlyHighAirways I didn't today... Must mean I need to take another trip!	@FlyHighAirways	I didn't today... Must mean I need to take another trip!	-0.390625	0.687500
3	@FlyHighAirways it's really aggressive to blast obnoxious "entertainment" in your guests' faces & they have little recourse	@FlyHighAirways	it's really aggressive to blast obnoxious "entertainment" in your guests' faces & they have little recourse	0.006250	0.350000

Figure 8.12: Tweets DataFrame with sentiment scores added

From the preceding figure, we can see that for each **tweet**, **Polarity**, and **Subjectivity** scores have been calculated.

14. To distinguish between the positive, negative, and neutral tweets, we need to add certain conditions. Consider tweets with polarity scores greater than **0.5** as positive, and tweets with polarity scores less than or equal to **-0.5** as negative. For neutral tweets, consider only those tweets that fall in the range of **-0.1** and **0.1**. Insert a new cell and add the following code to implement this:

```
positive_tweets = tweets[tweets.Polarity > 0.5]
negative_tweets = tweets[tweets.Polarity <= - 0.5]
neutral_tweets = tweets[ (tweets.Polarity > -0.1) \
                    & (tweets.Polarity < 0.1) ]
```

15. To view positive, negative, and neutral tweets, add the following code:

```
positive_tweets.head(15)
negative_tweets.head(15)
neutral_tweets
```

This displays the result of positive, negative, and neutral tweets. We have seen how to perform sentiment analysis using the **textblob** library. The following image shows the top four neutral tweets:

	tweet	At	tweet_preprocessed	Polarity	Subjectivity
0	@FlyHighAirways What @dhepburn said.	@FlyHighAirways	What said.	0.00000	0.00
1	@FlyHighAirways plus you've added commercials to the experience... tacky.	@FlyHighAirways	plus you've added commercials to the experience... tacky.	0.00000	0.00
3	@FlyHighAirways it's really aggressive to blast obnoxious "entertainment" in your guests' faces & they have little recourse	@FlyHighAirways	it's really aggressive to blast obnoxious "entertainment" in your guests' faces & they have little recourse	0.00625	0.35
10	@FlyHighAirways did you know that suicide is the second leading cause of death among teens 10-24	@FlyHighAirways	did you know that suicide is the second leading cause of death among teens 10-24	0.00000	0.00

Figure 8.13: Neutral tweets

NOTE

To access the source code for this specific section, please refer to https://packt.live/2XfcuIC.

You can also run this example online at https://packt.live/2DqDSfq.

ACTIVITY 8.02: TRAINING A SENTIMENT MODEL USING TFIDF AND LOGISTIC REGRESSION

Solution

1. Open a Jupyter Notebook.

2. Insert a new cell and add the following code to import the necessary libraries:

```
import pandas as pd
pd.set_option('display.max_colwidth', 200)
```

3. To load all three datasets, insert a new cell and add the following code:

```
DATA_DIR = 'data/sentiment_labelled_sentences/'


IMDB_DATA_FILE = DATA_DIR + 'imdb_labelled.txt'
YELP_DATA_FILE = DATA_DIR + 'yelp_labelled.txt'
AMAZON_DATA_FILE = DATA_DIR + 'amazon_cells_labelled.txt'


COLUMN_NAMES = ['Review', 'Sentiment']
yelp_reviews = pd.read_table(YELP_DATA_FILE, names=COLUMN_NAMES)
amazon_reviews = pd.read_table(AMAZON_DATA_FILE, \
                               names=COLUMN_NAMES)
imdb_reviews = pd.read_table(YELP_DATA_FILE, names=COLUMN_NAMES)
```

If we look at the code, even though the data comes from three different business domains, they are labeled and stored in the same format, which can help us to concatenate them together. This is the reason we can combine them to train our sentiment analysis model.

4. Now we concatenate the different datasets into one dataset using the **concat()** function. Insert a new cell and add the following code to implement this:

```
review_data = pd.concat([amazon_reviews, imdb_reviews, \
                         yelp_reviews])
```

Since we combined the data from three separate files, let's make use of the **sample()** function, which returns a random selection from the dataset. This will allow us to see the reviews from different files. Insert a new cell and add the following code to implement this:

```
review_data.sample(10)
```

The code generates the following output (only the top four reviews are displayed here):

	Review	Sentiment
233	Ordered a double cheeseburger & got a single patty that was falling apart (picture uploaded) Yeah, still sucks.	0
722	If you haven't choked in your own vomit by the end (by all the cheap drama and worthless dialogue) you've must have bored yourself to death with this waste of time.	0
95	We'll never go again.	0
873	My sister has one also and she loves it.	1

Figure 8.14: Output from calling the sample() function

5. To view the number of counts, add the following code:

```
review_data.Sentiment.value_counts()
```

6. Create a function named **clean()** and do some preprocessing. Basically, we need to remove unnecessary characters. Insert a new cell and add the following code to do this:

```
import re

def clean(text):
    text = re.sub(r'[\W]+', ' ', text.lower())
    text = text.replace('hadn t' , 'had not')\
              .replace('wasn t', 'was not')\
              .replace('didn t', 'did not')
    return text
```

In the preceding code snippet, first, the text is converted to lowercase and cleaned, and then keywords with apostrophes are converted into their original form.

7. Once the function is defined, we can clean and tokenize the text. It is a good practice to apply transformation functions on copies of our data unless you are really constrained with memory. Insert a new cell and add the following code to implement this:

```
review_model_data = review_data.copy()
review_model_data.Review = review_data.Review.apply(clean)
```

8. Now sample the data again to see what the processed text looks like. Add the following code in a new cell to implement this:

```
review_model_data.sample(10)
```

The code generates the following output (only the top four reviews are displayed here):

	Review	Sentiment
197	bad choice	0
533	it handles some tough issues with dignity and grace and of course has shocking spoiler here	1
57	not much seafood and like 5 strings of pasta at the bottom	0
176	if you see it you should probably just leave it on the shelf	0

Figure 8.15: Sample of 10 after cleaning the Review column

In the preceding figure, we can see that the text is converted to lowercase and only alphanumeric characters remain.

9. Now it is time to develop our model. We will use **TfidfVectorizer** to convert each review into a **TFIDF** vector. We will then use **LogisticRegression** to build a model. Insert a new cell and add the following code to import the necessary libraries:

```
from sklearn.feature_extraction.text import TfidfVectorizer
from sklearn.pipeline import Pipeline
from sklearn.model_selection import train_test_split
from sklearn.linear_model import LogisticRegression
```

Next, combine **TfidfVectorizer** and **LogisticRegression** in a **Pipeline** object. In order to do this, insert a new cell and add the following code:

```
tfidf = TfidfVectorizer()
log_reg = LogisticRegression()
log_tfidf = Pipeline([('vect', tfidf),\
                      ('clf', log_reg)])
```

10. Once the data is ready, split it into train and test sets. Split it into 70% for training and 30% for testing. This can be achieved with the help of the **train_test_split()** function. Insert a new cell and add the following code to implement this:

```
X_train, X_test, y_train, y_test = train_test_split\
                                    (review_model_data.Review, \
                                     review_model_data.Sentiment,\
                                     test_size=0.3, \
                                     random_state=42)
```

11. Fit the training data to the training pipeline with the help of the **fit()** function. Insert a new cell and add the following code to implement this:

```
log_tfidf.fit(X_train.values, y_train.values)
```

The code generates the following output:

```
Pipeline(memory=None,
     steps=[('vect', TfidfVectorizer(analyzer='word', binary=False, decode_error='strict',
         dtype=<class 'numpy.int64'>, encoding='utf-8', input='content',
         lowercase=True, max_df=1.0, max_features=None, min_df=1,
         ngram_range=(1, 1), norm='l2', preprocessor=None, smooth_idf=True,
     ...ty='l2', random_state=None, solver='liblinear', tol=0.0001,
         verbose=0, warm_start=False))])
```

Figure 8.16: Output from calling the fit() function on the training model

12. In order to check our model's **accuracy**, use the **score()** function. Insert a new cell and add the following code to implement this:

```
test_accuracy = log_tfidf.score(X_test.values, y_test.values)
'The model has a test accuracy of {:.0%}'.format(test_accuracy)
```

You should an output as follows:

```
'The model has a test accuracy of 81%'
```

As you can see from the preceding figure, our model has an accuracy of **81%**, which is pretty good for such a simple model.

13. The model is ready with an accuracy of **81%**. Now we can use it to predict the sentiment of sentences. Insert a new cell and add the following code to implement this:

```
log_tfidf.predict(['I loved this place', 'I hated this place'])
```

You should see an output like the following:

```
array([1, 0], dtype=int64)
```

In the preceding figure, we can see how our model predicts sentiment. For a positive test sentence, it returns a score of **1**. For a negative test sentence, it returns a score of **0**.

> **NOTE**
>
> To access the source code for this specific section, please refer to https://packt.live/3gto1M9.
>
> You can also run this example online at https://packt.live/2PowrIN.

INDEX

INDEX

U

X